Springer Monographs in Mathematics

This series publishes advanced monographs giving well-written presentations of the "state-of-the-art" in fields of mathematical research that have acquired the maturity needed for such a treatment. They are sufficiently self-contained to be accessible to more than just the intimate specialists of the subject, and sufficiently comprehensive to remain valuable references for many years. Besides the current state of knowledge in its field, an SMM volume should ideally describe its relevance to and interaction with neighbouring fields of mathematics, and give pointers to future directions of research.

Giovanni De Franceschi • Martin W. Liebeck •
Eamonn A. O'Brien

Conjugacy in Finite Classical Groups

Springer

Giovanni De Franceschi
Department of Mathematics
University of Auckland
Auckland, New Zealand

Martin W. Liebeck
Department of Mathematics
Imperial College London
London, UK

Eamonn A. O'Brien
Department of Mathematics
University of Auckland
Auckland, New Zealand

ISSN 1439-7382 ISSN 2196-9922 (electronic)
Springer Monographs in Mathematics
ISBN 978-3-031-86460-5 ISBN 978-3-031-86461-2 (eBook)
https://doi.org/10.1007/978-3-031-86461-2

This Springer imprint is published by the registered company Springer Nature Switzerland AG
The registered company address is: Gewerbestrasse 11, 6330 Cham, Switzerland

Preface

Let G be a classical group defined over a finite field. We consider the following fundamental problems concerning conjugacy in G:

- List a representative for each conjugacy class of G.
- Given $x \in G$, describe the centralizer of x in G, by giving its group structure and a generating set.
- Given $x, y \in G$, establish whether x and y are conjugate in G and, if so, then find explicit $z \in G$ such that $z^{-1}xz = y$.

We present comprehensive theoretical solutions to all three problems, and use our solutions to formulate practical algorithms. In parallel to our theoretical work, we have developed in MAGMA complete implementations of our algorithms. They form a critical component of various general algorithms in computational group theory—for example, computing character tables and solving conjugacy problems in arbitrary finite groups.

Auckland, New Zealand	Giovanni De Franceschi
London, UK	Martin W. Liebeck
Auckland, New Zealand	Eamonn A. O'Brien

Acknowledgments

Liebeck and O'Brien were supported by the Marsden Fund of New Zealand and by the 2020 and 2022 programs "Groups, representations and applications: new perspectives" at the Isaac Newton Institute, Cambridge.

Some of this work formed part of the PhD thesis of De Franceschi [9], supported by a University of Auckland PhD scholarship.

We thank Professor Donald Taylor of the University of Sydney for many helpful discussions about the project and related MAGMA code.

Contents

Chapter 1
Introduction and Background

This book is a study of conjugacy in finite classical groups: the general and special linear, symplectic, orthogonal and unitary groups over finite fields. In this chapter we introduce the three main conjugacy problems studied. In Sect. 1.1 we describe the problems and some of their history. In Sect. 1.2 we present both our strategy to solve them and a summary of our main results. Finally, in Sects. 1.3–1.6 we discuss our related algorithms and their performance, and the role they play in the solution of these problems for arbitrary finite groups.

1.1 The Main Problems

Let G be a classical group defined over a finite field. Our aim is to obtain a precise and explicit solution to each of the following closely related problems:

(1) List a representative for each conjugacy class of G.
(2) Given $x \in G$, describe the centralizer $C_G(x)$ of x in G, by giving its group structure and a generating set.
(3) Given $x, y \in G$, establish whether x and y are conjugate in G and, if so, then find explicit $z \in G$ such that $z^{-1}xz = y$.

We also aim to provide algorithms that implement our solution of each of the problems.

We achieve all of these aims in this work. First we present comprehensive theoretical solutions to all three problems, and then use our theoretical solutions to formulate practical algorithms to solve them. We have developed implementations of our algorithms in MAGMA [3]. These algorithms form a critical component of various general algorithms in computational group theory: as one example, algorithms for all three problems are vital in computing the character table of a

G. De Franceschi et al., *Conjugacy in Finite Classical Groups*, Springer Monographs in Mathematics, https://doi.org/10.1007/978-3-031-86461-2_1

classical group; as another, the algorithms are fundamental to the computational solution of conjugacy problems for arbitrary finite groups, as outlined in Sect. 1.6.

Various parts of the conjugacy problems (1)–(3) for finite classical groups have been much studied in the literature [4, 7, 12, 14, 20, 22, 25, 26, 32, 33]. One of the most significant such works is the 1963 paper of Wall [32]. In that paper, for a given classical isometry group G on a finite vector space V, the author gave criteria for an arbitrary element of the general linear group $\mathrm{GL}(V)$ to lie in G; he gave necessary and sufficient conditions for two elements of G to be conjugate; and he determined the total number of conjugacy classes in G and their sizes. The theory developed in [32] was set in the much broader context of classical groups over division rings, and does not lend itself well to our explicit and algorithmic requirements. The same comment applies to the approach of Carter [7], where the semisimple classes and centralizers in finite classical groups are determined using the theory of algebraic groups, and also to the work in [22], where the same is done for the unipotent classes. In summary, while various treatments in the literature give partial solutions to our problems, none is constructive: for example, the structure of the centralizer of an arbitrary element of a finite classical group is known, but nowhere is a method given to construct its generating set.

Nevertheless, we found various parts of the literature very useful for our approach to these problems. In particular, our work is influenced by the article of Milnor [25], as interpreted and developed by Britnell [4], and a short paper of Wall [33]; also we make full use of [14, 22] for representatives and centralizers of unipotent classes.

1.2 Background and Summary of Results

We first introduce some notation that we use throughout. If $A, A_1, \ldots, A_r$ are square matrices, then

$$\bigoplus_{i=1}^{r} A_i = A_1 \oplus \cdots \oplus A_r$$

denotes the block diagonal matrix with diagonal blocks A_i, and $A^{[s]}$ is the block diagonal matrix $A \oplus \cdots \oplus A$, where there are s blocks.

We begin in Chap. 2 by describing the conjugacy classes and centralizers in $\mathrm{GL}_n(q)$. Although the theory for this case is well known—see for example [5, §3.2], [16, Chap. 3], [24, §4.2] and [26]—we make a new contribution: an explicit description of the generators of the centralizer of an element. Moreover, the notation and concepts introduced in this chapter underpin our approach to the general problems, so we present the solution in detail.

Here is a brief summary. Let $q = p^a$, where p is prime and a is a positive integer, and let $F = \mathbb{F}_q$. Let $V = F^n$ be an n-dimensional vector space over F, and set $G = \mathrm{GL}(V) \cong \mathrm{GL}_n(F)$. For $x \in G$ with minimal polynomial $\prod_1^h f_i(t)^{e_i}$, where $f_1, \ldots, f_h \in F[t]$ are distinct monic irreducible polynomials, there is a decomposition

$$V = V_1 \oplus V_2 \oplus \cdots \oplus V_h, \tag{1.2.1}$$

where $V_i = \ker(f_i(x)^{e_i})$. Moreover for each i we can write $V_i = V_{i,1} \oplus \cdots \oplus V_{i,k_i}$, a sum of x-invariant subspaces such that x acts cyclically on each $V_{i,j}$ with minimal polynomial $f_i(t)^{e_{ij}}$, where $1 \leq e_{i1} \leq \cdots \leq e_{ik_i} = e_i$. The polynomials $f_i^{e_{ij}}$ are the *elementary divisors* of x. It is well known that elements of G are conjugate if and only if they have the same elementary divisors.

The choice of representative for each class is important for the solution of the conjugacy problems (2) and (3) from the previous section. If x and y are conjugate in G, then the easiest way to compute a conjugating element is to find matrices z_x and z_y such that $z_x^{-1} x z_x = J = z_y^{-1} y z_y$, where J is the representative for the conjugacy class of x and y. The desired conjugating element z is then $z_x z_y^{-1}$. Moreover, if $z^{-1} x z = y$, then $C_G(y) = z^{-1} C_G(x) z$; thus, to describe centralizers in G it is sufficient to describe the centralizers of the representatives of conjugacy classes, so it is convenient to choose a suitable canonical form. We use the *Jordan form* of x: it is a block diagonal matrix $J = \bigoplus_{i,j} J_{f_i, e_{ij}}$, where $J_{f_i, e_{ij}}$ is the matrix of the restriction of x to $V_{i,j}$ and

$$J_{f_i, e_{ij}} = \begin{pmatrix} C_i & \mathbb{I} & & \\ & \ddots & \ddots & \\ & & \ddots & \mathbb{I} \\ & & & C_i \end{pmatrix},$$

where C_i, the companion matrix of f_i, appears e_{ij} times. The matrix $J_{f_i, e_{ij}}$ is the *Jordan block* of order e_{ij} relative to f_i. This is a generalization of the canonical Jordan form (defined in [16, §6.6] only for matrices whose characteristic polynomial splits into linear factors), and is used also by Macdonald [24, §4.2] and Murray [26].

Recall that $x \in G \cong \mathrm{GL}_n(q)$ is *semisimple* if its order is coprime to p, and x is *unipotent* if its order is a power of p. If x is semisimple, then $e_{ij} = 1$ for all i, j; if x is unipotent, then $h = 1$ and $f_1(t) = t - 1$. The *Jordan decomposition* states that every $x \in G$ can be written uniquely as

$$x = su = us, \tag{1.2.2}$$

where s is semisimple and u is unipotent. In particular, $u \in C_G(s)$. Two important consequences are the following:

- Let $x, y \in G$ and let $x = s_x u_x$ and $y = s_y u_y$ be their Jordan decompositions. Then x and y are conjugate in G if and only if s_x and s_y are conjugate in G and $z^{-1} u_x z$ and u_y are conjugate in $C_G(s_y)$, where $z \in G$ is such that $z^{-1} s_x z = s_y$.

- If $x \in G$ and $x = su$ is its Jordan decomposition, then $C_G(x) = C_G(s) \cap C_G(u) = C_{C_G(s)}(u)$.

Hence, to solve conjugacy problems in G, it is usually sufficient to solve them in G for semisimple elements s and in $C_G(s)$ for unipotent elements. One additional simplification is the following. Let $X_i = \bigoplus_{j=1}^{k_i} J_{f_i,e_{ij}}$ be the matrix of the restriction of x to V_i. If $y \in C_G(J)$, then $V_i y \subseteq V_i$, so y has block diagonal form $y = \bigoplus_{i=1}^{m} y_i$, with $y_i X_i = X_i y_i$ for every i. Hence, we can solve the centralizer problem separately on each V_i.

Thus we may replace J by a single X_i. Having done this, let $J = J_u J_s$ be the Jordan decomposition of J. Then $J_s = C^{[m]} = C \oplus \cdots \oplus C$, with C the companion matrix of an irreducible $f \in F[t]$ of degree d, and $C_G(J_s) \cong \mathrm{GL}_m(q^d)$ (see Sect. 2.2.1). Moreover, J_u is a unipotent element of this centralizer. In [26, §2.3], Murray describes the centralizer in a general linear group of a unipotent element. In Sect. 2.2.2 we expand this description and obtain an algorithm which lists an explicit generating set for such a centralizer. For completeness, we discuss in Sect. 2.3 classes and centralizers in the special linear group $\mathrm{SL}_n(q)$.

In the rest of the work, from Chap. 3 onwards, we consider the other classical groups—namely, the symplectic, orthogonal and unitary groups. We define these groups and record some of their basic properties in Sect. 3.1. We then establish which conjugacy classes in $\mathrm{GL}_n(F)$ contain elements in these groups of isometries. This problem was considered by Britnell [4, Chap. 5], Milnor [25] and Wall [32, §2.6, §3.7]. We mainly refer to [4, Chap. 5], where Britnell solves the membership problem in symplectic and orthogonal groups of odd characteristic. In Sect. 3.2 we present his results, and extend them to hermitian and quadratic forms (including characteristic 2). The main result is Theorem 3.2.6, which gives necessary and sufficient criteria for $x \in \mathrm{GL}(V)$ to preserve a non-degenerate alternating, hermitian or quadratic form on V, in terms of the elementary divisors of x.

As in the case of the general linear group, the key to analysing conjugacy in the classical groups is the Jordan decomposition. As noted already, if G is a classical group over a finite field F, then to solve the conjugacy problems in G it is sufficient to solve it in G for semisimple elements s and in $C_G(s)$ for unipotent elements. The structure of $C_G(s)$ is described in Chap. 6: as summarised in (1.2.7) below, when G is a full isometry group, $C_G(s)$ is a direct product of classical groups of smaller dimension over extension fields of F. Hence a key component of our work is the solution of the conjugacy problems for unipotent elements in classical groups. This is carried out in Chaps. 4 and 5. Let p be the characteristic of the field F. If G is symplectic or orthogonal and $p = 2$, then p is *bad* for G; otherwise p is *good* for G. The three conjugacy problems are much more complicated to solve for unipotent classes of classical groups in bad characteristic than in good; they are solved in Chaps. 4 and 5 for good and bad characteristic respectively. In both cases, explicit representatives for the unipotent classes in G are given by [14]; and the structures and orders of the centralizers can be read off from [22]. However, computing generators for the centralizers and solving the other conjugacy problems presents new challenges.

Here is a brief description of our approach, first for the case of good characteristic. To keep the exposition reasonably brief, we restrict the discussion to the symplectic group $G = \mathrm{Sp}(V) \cong \mathrm{Sp}_{2n}(q)$ (with q odd). Let J_i denote an $i \times i$ unipotent Jordan block. It is well known that a block diagonal matrix $\bigoplus_i J_i^{[n_i]}$ is the Jordan form of a unipotent element of G if and only if n_i is even for all odd values of i—so all odd-sized Jordan blocks appear with even multiplicities. In Sect. 4.1.1 we list new representatives for the unipotent classes of G. These are labelled

$$u = \bigoplus_{i=1}^{r} (V_{\beta_i}(2k_i) \oplus V_1(2k_i)^{[a_i-1]}) \oplus \bigoplus_{i=1}^{s} W(2l_i+1)^{[b_i]}, \tag{1.2.3}$$

where each summand $V_1(2k_i)$ or $V_{\beta_i}(2k_i)$ is a single Jordan block of size $2k_i$, and each summand $W(2l_i + 1)$ comprises two Jordan blocks of size $2l_i + 1$. Each summand in (1.2.3) corresponds to a nilpotent element of the symplectic Lie algebra, via the *Cayley map*. This is a G-equivariant bijection from the set of nilpotent elements of the Lie algebra to the set of unipotent elements of G: a nilpotent element e is mapped to the unipotent element $(1 - e)(1 + e)^{-1}$ of G. For example, a summand $V_1(2k)$ corresponds to the nilpotent element

$$\begin{pmatrix} 0 & 1 & & & \\ & 0 & 1 & & \\ & & \ddots & & \\ & & & 0 & 1 \\ & & & & 0 \end{pmatrix}$$

(matrices with respect to a standard symplectic basis). There is a 1-dimensional torus naturally associated with this nilpotent element, namely $\{T(c) : c \in \bar{F}^*\}$, where $\bar{F}$ is the algebraic closure of F and $T(c) = \mathrm{diag}(c^{-(2k-1)}, c^{-(2k-3)}, \ldots, c^{2k-1})$ (so that $e\,T(c) = c^2 e$). We can define a 1-dimensional torus T in this way corresponding to a unipotent element as in (1.2.3), and the stabilizer in G of the flag of V defined by sums of T-weight spaces for decreasing weights is a parabolic subgroup P. Write $P = QL$, where Q is the unipotent radical and $L = C_G(T)$ is a Levi factor. Then $C_G(u) \leq P$ (see Theorem 4.1.2); indeed, $C_G(u) = C_Q(u) \rtimes C_L(u)$. We compute generators for both factors in Sect. 4.2.1, solving the centralizer problem.

Continuing the discussion of symplectic groups in good characteristic, the problem of deciding conjugacy is fairly straightforward: given unipotent $g \in G$, we must decide which representative of the form (1.2.3) it is conjugate to. This is determined by the Jordan form of g, apart from computing the values of the parameters β_i, for which an algorithm is given in Sect. 4.3.1. More challenging is the problem of finding a conjugating element: for example, given $g \in G = \mathrm{Sp}_{2k}(q)$ that is conjugate to a block $u = V_1(2k)$, compute $y \in G$ such that $g^y = u$. A key observation is that if e is the associated nilpotent element given above, then V has a basis

$$v,\ ve,\ ve^2, \ldots, ve^{2k-1}, \tag{1.2.4}$$

and this is a standard basis for the symplectic form. Consider the nilpotent element $f := (1-g)(1+g)^{-1}$. We seek $w \in V$ such that $w, wf, wf^2, \ldots, wf^{2k-1}$ is a basis of V with symplectic form values matching those of (1.2.4). Then the map y sending $wf^i \mapsto ve^i$ for all i is the desired conjugating element. An algorithm to compute such a vector w is given in Sect. 4.4.1. Carrying this out for each block in an arbitrary unipotent element (1.2.3) leads to the solution of the conjugating element problem in general.

As remarked above, the analysis of unipotent classes in bad characteristic is more complicated than in good characteristic. Let $G = \mathrm{Sp}(V)$ or $\mathrm{O}(V)$ in characteristic 2. Representatives for the unipotent conjugacy classes were written down in [14], and take the form

$$u = \bigoplus_i W(m_i)^{[c_i]} \oplus \bigoplus_j V(2k_j)^{[d_j]} \oplus \bigoplus_r W_\alpha(m_r') \oplus \bigoplus_s V_\alpha(2k_s'), \tag{1.2.5}$$

where the parameters m_i, m_r', k_j, k_s' satisfy various numerical conditions (see (i)–(v) given after (5.1.1)), and α is a fixed element of F such that the quadratic $x^2+x+\alpha$ is irreducible in $F[x]$. Again, each summand $V(2k_i)$ or $V_\alpha(2k_i)$ is a single Jordan block of size $2k_i$, and each summand $W(m_i)$ or $W_\alpha(m_i)$ comprises two Jordan blocks of size m_i. In characteristic 2 there is no Cayley map. Nevertheless, we can associate weights to the basis vectors for each summand, and use these to define a canonical parabolic subgroup P corresponding to the class representative u. A deep result of Clarke and Premet [8] implies that $C_G(u) \le P$. We do not have a factorization $C_G(u) = C_Q(u) \rtimes C_L(u)$ as in the good characteristic case—indeed, $C_P(u)$ can be a nonsplit extension of $C_Q(u)$. However, we can work in P to compute generators for $C_G(u)$; this is done in Sect. 5.2.

The question of deciding conjugacy is also more complex for the bad characteristic case, and is solved in Sect. 5.3. The problem of computing conjugating elements is the most challenging, and is solved in Sect. 5.4. As an illustration, consider a single block $u = V_\alpha(2k)$: suppose $g \in G = \mathrm{Sp}_{2k}(q)$ (for $q = 2^a$) is conjugate to u; we wish to compute $y \in G$ such that $g^y = u$. To do this, we would like to write down a standard basis in similar fashion to (1.2.4) above, in terms of a single vector v and a nilpotent operator e. It turns out that this can be done if we use three nilpotent operators, rather than one: let $f = 1 + u$ and $e = f + f^2 + \cdots + f^{2k-1}$. We show that there is an additional nilpotent operator h such that the following sequence of vectors is a standard basis for the symplectic form:

$$v,\ ve, \ldots, ve^{k-2},\ v(e^{k-1}+h),\ v(e^{k-1}+h)f, \ldots, v(e^{k-1}+h)f^k. \tag{1.2.6}$$

There is a polynomial $p(x)$ of degree $2k - 1$ such that $h = p(f)$—we cannot write $p(x)$ down explicitly, but it can be computed by machine for any given k and q, which is enough for our algorithm. Now we take a similar approach to that for good characteristic. We let $f_0 = 1 + g$, $e_0 = f_0 + \cdots + f_0^{2k-1}$, $h_0 = p(f_0)$, and seek $w \in V$ such that

$$w,\ we_0, \ldots, we_0^{k-2},\ w(e_0^{k-1} + h_0),\ w(e_0^{k-1} + h_0)f_0, \ldots, w(e_0^{k-1} + h_0)f_0^k$$

is a basis of V with symplectic form values matching those of (1.2.6). An algorithm to compute such a vector w is given in Sect. 5.4. Once again, applying this procedure for each block in an arbitrary unipotent element (1.2.5) leads to the solution of the conjugating element problem in general.

Having dealt with the unipotent classes, in Chap. 6 we consider the semisimple classes. Our starting point is a brief paper of Wall [33] where the semisimple conjugacy classes in symplectic groups of odd characteristic are classified. In Sect. 6.1 we extend this work to all sesquilinear and quadratic forms in all positive characteristics. The structure of the centralizer appears in [5, Chap. 3] and [12, §1]. Here we briefly discuss our description of the centralizer of a semisimple element for the symplectic and orthogonal groups, given in Theorem 6.1.2. For a monic polynomial $f(t) = t^d + a_{d-1}t^{d-1} + \cdots + a_0 \in \mathbb{F}_q[t]$ with $a_0 \neq 0$, define the *dual polynomial* $f^*(t) = a_0^{-1}t^d f(t^{-1})$, another monic polynomial of degree d. Following [12, §1], define

$$\Phi_1 := \{f : f \in F[t] \mid f = f^* \text{ monic irreducible, } \deg f = 1\};$$

$$\Phi_2 := \{f : f \in F[t] \mid f = gg^*,\ g \neq g^*,\ g \text{ monic irreducible}\};$$

$$\Phi_3 := \{f : f \in F[t] \mid f = f^* \text{ monic irreducible, } \deg f > 1\}.$$

Define $\Phi := \Phi_1 \cup \Phi_2 \cup \Phi_3$. The elementary divisors of every semisimple $s \in G = I_n(q)$ (a symplectic or orthogonal isometry group) are either in $\Phi_1 \cup \Phi_3$, or appear in pairs g, g^*, where $gg^* \in \Phi_2$. The polynomials in Φ are the *generalized* elementary divisors of s. For each $f \in \Phi$, let m_f be its multiplicity as a generalized elementary divisor of s, and let $d_f = \frac{1}{2}\deg(f)$. Then

$$C_G(s) = \prod_{f\in\Phi_1} I_{m_f}(q) \times \prod_{f\in\Phi_2} \mathrm{GL}_{m_f}(q^{d_f}) \times \prod_{f\in\Phi_3} \mathrm{GU}_{m_f}(q^{d_f}). \qquad (1.2.7)$$

An algorithm to compute a generating set for this centralizer is given in Sect. 6.3.

In Chap. 7 we use the theory developed for semisimple and unipotent classes to solve the three conjugacy problems in the general case. The listing of conjugacy classes in G proceeds as follows. We first list all semisimple classes; each is identified by a pair (s, B), where s and B are the matrices of the representative and of the form respectively. If $h_1, \ldots, h_r \in \Phi$ are the distinct generalized elementary divisors of s, then $s = \bigoplus_{i=1}^r s_i$ and $B = \bigoplus_{i=1}^r B_i$, where s_i and B_i are the restrictions of s and B to the generalized eigenspace relative to h_i. If G_i is the isometry group of the form B_i, then $C_{G_i}(s_i)$ is one of the factors listed in (1.2.7). Now multiply s by all matrices of the form $u = \bigoplus_{i=1}^{r} u_i$, where u_i runs over the set of representatives for unipotent classes in $C_i := C_{G_i}(s_i)$. The set of pairs (su, B) built in this way is a complete set of representatives for conjugacy classes in G (see Theorem 7.1.1). For every (su, B), the centralizer of su in G is $\prod_i C_{C_i}(u_i)$ (see Theorem 7.1.1). An algorithm to compute a generating set for this centralizer is given in Sect. 7.2. The conjugacy problem is addressed in Sect. 7.3. Finally, given conjugate $x, y \in G$, this theory is used in Sect. 7.4 to compute $z \in G$ such that $x^z = y$.

1.3 Providing Electronic Access to the Results

We have developed in MAGMA [3] implementations of our algorithms to solve the conjugacy problems (1)–(3) for classical groups. The resulting code is available publicly at [10] and is also distributed as part of MAGMA.

By default, our conjugacy class representatives are returned as elements of the *standard copy* of the classical group $\mathcal{C} = \mathrm{SL}(V), \mathrm{SU}(V), \mathrm{Sp}(V), \mathrm{SO}(V)$, or $\Omega(V)$, as defined in MAGMA; classes in the corresponding isometry groups are also determined. The functions to list conjugacy classes in $\mathcal{C}$ return a sequence of triples $\langle |r|, |r^{\mathcal{C}}|, r\rangle$, where r is a representative for a conjugacy class of $\mathcal{C}$. The semisimple and unipotent classes can be constructed independently. We provide a function that, given semisimple $s \in \mathcal{C}$, returns representatives for all conjugacy classes of $\mathcal{C}$ having s as semisimple part. As discussed in Sect. 6.3, by applying the algorithm of [34], the data is readily translated to any other natural copy of $\mathcal{C}$.

Table 1.1 records the CPU time in seconds to construct the representatives for conjugacy classes of some classical groups $\mathcal{C}$; it also records the total times taken to

Table 1.1 Some calculations in classical groups

$\mathcal{C}$	Number of classes	Setup	Centralizer	Conjugation
$\Omega_{10}^{+}(4)$	1543	2	12	14
$\Omega_{10}^{-}(4)$	1593	3	13	15
$\Omega_{10}^{+}(9)$	36,177	24	148	154
$\Omega_{10}^{-}(9)$	36,205	29	160	177
$\Omega_{11}(5)$	3771	4	38	37
$\Omega_{11}(9)$	44,933	57	319	316
$\mathrm{Sp}_{10}(4)$	2170	2	13	11
$\mathrm{Sp}_{10}(9)$	107,992	46	605	519
$\mathrm{SU}_{10}(2)$	2340	1	45	30
$\mathrm{SU}_{10}(3)$	41,218	18	537	392
$\mathrm{SL}_{10}(4)$	349,420	46	644	128

construct the centralizer in $\mathcal{C}$ of a random element from each class, and to construct an element of $\mathcal{C}$ which conjugates between two random elements from each class. All calculations were carried out using MAGMA 2.28-13 on a 2.6 GHz machine. The relevant commands are listed at [10].

1.4 The Complexity of Our Algorithms

We comment briefly on the theoretical complexity of our solutions. Of course, many tasks are solved by writing down explicit solutions. Most of our remaining tasks for elements of a classical group $\mathcal{C}$ of degree n defined over a field of size q are solved in time $O(n^3 \log q)$. These include constructing Jordan forms, and constructing and factorising minimal polynomials; see for example [13, 17, 27]. We identify two exceptions.

- To construct the centralizer in $\mathcal{C}$ of a unipotent element x, we construct an isomorphic copy P of a unipotent subgroup of $\mathcal{C}$, describe P by a power-conjugate presentation, and construct the centralizer in P of x. While calculations using such presentations are practically very efficient, they rely on "collection", an algorithm not known to run in polynomial time (see [17, Chap. 9]).
- If x and y are unipotent conjugate elements of a classical group $\mathcal{C}$ of bad characteristic, then we write down a system of quadratic equations to construct $z \in \mathcal{C}$ such that $x^z = y$. This system is solved using a Gröbner basis algorithm (see [1, Chap. 1]). While our systems are solved readily in practice, this algorithm is not known to run in polynomial time.

1.5 Conjugacy in Central Quotients

As above, let $\mathcal{C}$ be a standard copy of $\mathrm{SL}(V)$, $\mathrm{SU}(V)$, $\mathrm{Sp}(V)$, $\mathrm{SO}(V)$, or $\Omega(V)$. Having solved the conjugacy problems for $\mathcal{C}$, we can solve them for any central quotient $\bar{\mathcal{C}} = \mathcal{C}/Z$ in an arbitrary representation, including the corresponding finite simple classical group. For such a group $\bar{\mathcal{C}}$, there is a constructive recognition algorithm that provides an explicit surjective homomorphism $\phi : \mathcal{C} \to \bar{\mathcal{C}}$; for details of such algorithms, see for example [11]. Using this machinery we can solve the conjugacy problems for $\bar{\mathcal{C}}$ as follows.

(1) Class representatives: Let $(g_i)_{i \in I}$ be a set of class representatives in $\mathcal{C}$. Define an equivalence relation on these by setting $g_i \sim g_j$ if and only if g_j is $\mathcal{C}$-conjugate to $g_i z$ for some $z \in Z$. A set of equivalence class representatives $(g_{i_j})_{j \in J}$ can be selected using the solution to the conjugacy problem in $\mathcal{C}$. Then $(\phi(g_{i_j}))_{j \in J}$ is a set of class representatives in $\bar{\mathcal{C}}$.

(2) Centralizers: Let $\phi(g) \in \bar{\mathcal{C}}$. For each $z \in Z$ such that g is $\mathcal{C}$-conjugate to gz, find $x_z \in \mathcal{C}$ such that $g^{x_z} = gz$. Then the centralizer of $\phi(g)$ in $\bar{\mathcal{C}}$ is generated by the image under ϕ of $C_{\mathcal{C}}(g)$, together with the images of these conjugating elements x_z.
(3) Conjugacy problem and conjugating element: $\phi(g)$ and $\phi(h)$ are conjugate in $\bar{\mathcal{C}}$ if and only if g is $\mathcal{C}$-conjugate to hz for some $z \in Z$. If $\phi(g)$ and $\phi(h)$ are $\bar{\mathcal{C}}$-conjugate, then a conjugating element is $\phi(x)$ where $g^x = hz$.

Our implementations of the resulting algorithms in MAGMA are available at [10].

1.6 Conjugacy in Arbitrary Finite Groups

We expect that our solution to the conjugacy problems for classical groups can assist in their solution for an arbitrary finite group G. Here is a brief sketch of how this is done in practice. Existing algorithms follow the "soluble radical" model (see [17, Chap. 10]). An efficient practical algorithm to construct the necessary data structure for this model is described in [2] and is available in MAGMA. Its output is a characteristic series for G:

$$1 \leqslant L \leqslant S \leqslant P \leqslant G,$$

where

- L is the solvable radical of G;
- S/L is the socle of G/L with $S/L \cong \prod_i T_i^{d_i}$, where the T_is are non-abelian, pairwise non-isomorphic simple groups;
- $P/S \leqslant \prod_i \mathrm{Out}(T_i)^{d_i}$ is solvable;
- $G/P \leqslant \prod_i \mathrm{Sym}(d_i)$.

Observe that we have a monomorphism from G/L into the direct product $\prod_i W_i$ where $W_i = \mathrm{Aut}(T_i) \wr \mathrm{Sym}(d_i)$.

- The problem to extend the solutions from $\mathrm{Aut}(T_i)$ to $\mathrm{Aut}(T_i) \wr \mathrm{Sym}(d_i)$, and then to G/L, was solved by Cannon and Holt [6] and Hulpke [18].
- Since L is solvable, there exists a series

  $$L = N_1 \rhd N_2 \rhd \cdots \rhd N_r = 1$$

  with N_i/N_{i+1} elementary abelian. For every i, the solution of the problem in G/N_{i+1} can be obtained from that in G/N_i. This procedure is described in [18, 19] for conjugacy classes and [17, §8.8] for centralizers.

Hence the conjugacy problems in arbitrary finite groups reduce to their solutions for finite *almost simple* groups.

1.7 Notation

Our standard references for notation are [21] and [30]. See also the notation index. Symbols used include the following.

$q = p^a$	for prime number p and positive integer a;
$\mathbb{F}_q$	finite field of size q;
F^*, $\mathbb{F}_q^*$	unit group of F, $\mathbb{F}_q$ respectively;
F^{*2}, $\mathbb{F}_q^{*2}$	subgroups of squares in F^*, $\mathbb{F}_q^*$ respectively;
$F[t]$	polynomial ring with coefficients in F;
$M_n(F)$	algebra of $n \times n$ matrices with entries in F;
$\mathrm{GL}_n(F)$	general linear group defined on the vector space F^n;
$V_n(q)$	vector space of dimension n over $\mathbb{F}_q$;
$\mathrm{GL}_n(q)$	general linear group defined on the vector space $\mathbb{F}_q^n$;
$\mathrm{GL}(V)$	general linear group defined on the vector space V;
g^U	action of matrix g on space U;
$\mathbb{O}$, $\mathbb{O}_n$	zero matrix (dimension not specified), $n \times n$ zero matrix;
$\mathbb{I}$, $\mathbb{I}_n$	identity matrix (dimension not specified), $n \times n$ identity matrix;
$A \oplus B$	block diagonal matrix $\begin{pmatrix} A & \mathbb{O} \\ \mathbb{O} & B \end{pmatrix}$;
$\bigoplus M_i$	block diagonal sum of matrices M_i;
$M^{[r]}$	block diagonal sum $M \oplus \cdots \oplus M$ (r blocks);
X^{t}	transpose of matrix X;
$a \mapsto \overline{a}$	automorphism of $\mathbb{F}_q$ of order 1 or 2;
$\overline{X}$	matrix $(\overline{a_{ij}})$, where $X = (a_{ij})$;
X^*	matrix $\overline{X}^{\mathrm{t}}$;
$f(t)^*$	polynomial $\overline{a}_0^{-1} \sum_{i=0}^{n} \overline{a}_{n-i} t^i$, where $f(t) = \sum_{i=0}^{n} a_i t^i$, and $a_0 \neq 0$;
$\deg f$	degree of polynomial f;
$\mathrm{rk}\, X$	rank of X;
δ_{ij}	Kronecker delta: $\delta_{ij} = 1$ if $i = j$, 0 otherwise;
$V \downarrow X$	restriction of a G-module V to X, an element or subgroup of G;
$[n]$	group of order n;
$\mathrm{diag}(\alpha_1, \ldots, \alpha_n)$	diagonal matrix with diagonal entries $\alpha_1, \ldots, \alpha_n$.

Chapter 2
General and Special Linear Groups

In this chapter we give a complete description of conjugacy classes and centralizers in the general and special linear groups. Although the theory for this case is well known, we make a new contribution: an explicit generating set for the centralizer of an arbitrary element.

2.1 Conjugacy Classes in $\mathrm{GL}_n(q)$

Let $q = p^a$, where p is prime and a is a positive integer, and let $F = \mathbb{F}_q$. Let V be an n-dimensional vector space over F and let $x \in \mathrm{GL}(V) \cong \mathrm{GL}_n(q)$. Let $f_1(t)^{e_1} \cdots f_h(t)^{e_h}$ be the minimal polynomial of x, where $f_1, \ldots, f_h \in F[t]$ are distinct monic irreducible polynomials. We write

$$V = V_1 \oplus \cdots \oplus V_h,$$

where $V_i = \ker(f_i(x)^{e_i})$ is the *generalized eigenspace* corresponding to $f_i(t)$ for every i. By [16, 4.5.1], every V_i can be written as a direct sum of x-invariant subspaces

$$V_i = V_{i,1} \oplus \cdots \oplus V_{i,k_i}, \tag{2.1.1}$$

where x acts cyclically on $V_{i,j}$ with minimal polynomial $f_i(t)^{e_{ij}}$, and $1 \leq e_{i1} \leq \cdots \leq e_{ik_i} = e_i$. The polynomials $f_1(t)^{e_{11}}, \ldots, f_1(t)^{e_{1k_1}}, \ldots, f_h(t)^{e_{h1}}, \ldots, f_h(t)^{e_{hk_h}}$ are the *elementary divisors* of x. If g is a power of an irreducible polynomial, then the *multiplicity* of g as an elementary divisor of x is the number of times g appears in the list of elementary divisors of x. Since the decomposition in (2.1.1) is unique up to rearranging the factors, the list of elementary divisors of x

G. De Franceschi et al., *Conjugacy in Finite Classical Groups*, Springer Monographs in Mathematics, https://doi.org/10.1007/978-3-031-86461-2_2

is well-defined. Elements of $\mathrm{GL}(V)$ are conjugate if and only if they have the same elementary divisors; see for example [16, 6.7.3].

As a representative for each conjugacy class of $\mathrm{GL}(V)$, we choose the Jordan form, defined as follows. For every monic irreducible polynomial f of degree d and positive integer e, the *Jordan block* $J_{f,e}$ of order e relative to f is the block matrix

$$J_{f,e} = \begin{pmatrix} C & \mathbb{I} & & \\ & \ddots & \ddots & \\ & & \ddots & \mathbb{I} \\ & & & C \end{pmatrix},$$

where C, the companion matrix of f, appears e times, and $\mathbb{I}$ is the $d \times d$ identity matrix. A Jordan block is *unipotent* if $f(t) = t - 1$. For every $V_{i,j}$ as in (2.1.1) there is a basis such that the matrix of the restriction of x to $V_{i,j}$ is the Jordan block $J_{f_i,e_{ij}}$. Hence, there exists a basis of V such that the matrix of x is a diagonal join of Jordan blocks. This matrix is the *Jordan form* of x.

We comment briefly on problems (1) and (3) of Sect. 1.1: we take as conjugacy class representatives in $\mathrm{GL}_n(q)$ the list of distinct Jordan forms of its elements; and for $x \in \mathrm{GL}_n(q)$, an element that conjugates x to its Jordan form is determined by a Jordan basis algorithm (see for example [29]). Hence we focus on the construction of the centralizer of x in $\mathrm{GL}_n(q)$.

2.2 Centralizers in $\mathrm{GL}_n(q)$

In this section we record how to construct the centralizer of $x \in \mathrm{GL}(V) \cong \mathrm{GL}_n(q)$. The results are well known, except for the generation of centralizers of unipotent elements; our work is motivated by that of Murray [26].

Let $x \in G := \mathrm{GL}(V)$ be in Jordan form, with minimal polynomial $f_1(t)^{e_1} \cdots f_h(t)^{e_h}$ and associated generalized eigenspaces V_i. Every element of $C_G(x)$ fixes each V_i. Hence, if $x = \bigoplus_{i=1}^h x_i$, where x_i is the restriction of x to V_i, then

$$C_{\mathrm{GL}(V)}(x) = \prod_{i=1}^{h} C_{\mathrm{GL}(V_i)}(x_i). \tag{2.2.1}$$

Recall the *Jordan decomposition*: if $x \in G$, then $x = su = us$ for unique semisimple s and unipotent u. If x is a diagonal join of Jordan blocks, then s and u are the diagonal joins of the semisimple and the unipotent parts of each block. To compute $C_G(x)$, we first compute $C_G(s)$ and then, using the fact that $u \in C_G(s)$, compute $C_{C_G(s)}(u)$.

2.2.1 Centralizer of a Semisimple Element

Let $x \in G := \mathrm{GL}_n(q)$ be semisimple. By (2.2.1), we can assume that x has a unique elementary divisor $f(t) \in \mathbb{F}_q[t]$, which is irreducible.

If $\deg f = 1$, then x is a scalar matrix, so $C_G(x) = \mathrm{GL}_n(q)$.

Now suppose that $f(t) \in \mathbb{F}_q[t]$ is irreducible of degree $r > 1$. Let $E = \mathbb{F}_q[t]/(f)$ be the splitting field of f over $\mathbb{F}_q$ and let $\alpha \in E$ be a root of f. Every element of E can be written as $\phi(\alpha)$ for some polynomial $\phi(t) \in \mathbb{F}_q[t]$ of degree smaller than r. For every positive integer m, there is a canonical embedding of $\mathrm{GL}_m(E) = \mathrm{GL}_m(q^r)$ into $\mathrm{GL}_{mr}(q)$ sending the matrix $(\phi_{ij}(\alpha))$ into the block matrix $(\phi_{ij}(C))$, where C is the companion matrix of f (see [5, 2.1.4]).

Now assume x is semisimple with a unique elementary divisor $f(t)$ of degree $r > 1$ and multiplicity m. We can suppose that the matrix of x is a diagonal join of m copies of C, the companion matrix of f. Then x is the embedding into $\mathrm{GL}_{mr}(q)$ of the scalar matrix $\alpha\mathbb{I}_m \in \mathrm{GL}_m(E)$. Every matrix of $\mathrm{GL}_m(E)$ commutes with $\alpha\mathbb{I}_m$, so its embedding into $\mathrm{GL}_{mr}(q)$ commutes with x. On the other hand, these are the only matrices in $\mathrm{GL}_{mr}(q)$ that commute with x (see [5, Lemma 3.1.9]). Thus we have proved the following.

Proposition 2.2.1 *Let $x \in \mathrm{GL}_n(q)$ be semisimple with characteristic polynomial $\prod_{i=1}^{h} f_i(t)^{e_i}$, where $f_1, \ldots, f_h \in \mathbb{F}_q[t]$ are distinct irreducibles and $\deg f_i = r_i$. Then*

$$C_{\mathrm{GL}_n(q)}(x) \cong \prod_{i=1}^{h} \mathrm{GL}_{e_i}(q^{r_i}).$$

2.2.2 Centralizer of a Unipotent Element

Let $x \in G := \mathrm{GL}_n(q)$ be unipotent of order p^b. The structure of $C_G(x)$ is well known; our new contribution is to describe an explicit generating set in Sect. 2.2.3. Following the approach of [26], it is convenient to work in the matrix algebra $M = M_n(\mathbb{F}_q)$ and find the centralizer $C_M(x)$ of x in M. The centralizer of x in $\mathrm{GL}_n(q)$ is the set of invertible elements of $C_M(x)$.

Since the unique eigenvalue of x is 1, the Jordan form of x is

$$\begin{pmatrix} J_{\lambda_1} & & & \\ & J_{\lambda_2} & & \\ & & \ddots & \\ & & & J_{\lambda_k} \end{pmatrix},$$

where J_{λ_i} is the unipotent Jordan block of dimension λ_i and $\lambda_1 + \cdots + \lambda_k = n$. We suppose $\lambda_1 \leq \lambda_2 \leq \cdots \leq \lambda_k$ and denote this partition of n by λ. Let x be this Jordan form.

Take an element y centralizing x and write the matrix of y as

$$\begin{pmatrix} B_{11} & \cdots & B_{1k} \\ \vdots & \ddots & \vdots \\ B_{k1} & \cdots & B_{kk} \end{pmatrix},$$

where B_{ij} is a block of dimension $\lambda_i \times \lambda_j$ for every i, j.

The condition $xy = yx$ implies $J_{\lambda_i} B_{ij} = B_{ij} J_{\lambda_j}$ for every $1 \leq i, j \leq k$. Write $B_{ij} = (b_{u,v})$ for $b_{u,v} \in \mathbb{F}_q$, where $0 \leq u \leq \lambda_i - 1$ and $0 \leq v \leq \lambda_j - 1$. A simple computation shows that $xy = yx$ is equivalent to $b_{u,v} = b_{u+1,v+1}$ for every u, v; also $b_{u,v} = 0$ if either $\lambda_i \leq \lambda_j$ and $u \geq v - (\lambda_j - \lambda_i - 1)$, or $\lambda_i > \lambda_j$ and $u \geq v + 1$. Hence B_{ij} is an upper triangular rectangular matrix with constant upper diagonals; it is either

$$\begin{pmatrix} \mathbb{O} & \begin{matrix} b_{\lambda_j-\lambda_i} & b_{\lambda_j-\lambda_i+1} & \cdots & b_{\lambda_j-1} \\ 0 & b_{\lambda_j-\lambda_i} & \ddots & \vdots \\ \vdots & \ddots & \ddots & b_{\lambda_j-\lambda_i+1} \\ 0 & \cdots & 0 & b_{\lambda_j-\lambda_i} \end{matrix} \end{pmatrix} \text{ or } \begin{pmatrix} \begin{matrix} b_0 & b_1 & \cdots & b_{\lambda_j-1} \\ 0 & b_0 & \ddots & \vdots \\ \vdots & \ddots & \ddots & b_1 \\ 0 & \cdots & 0 & b_0 \end{matrix} \\ \mathbb{O} \end{pmatrix}. \quad (2.2.2)$$

Denote by $X^a_{c\times d}$ the $c \times d$ matrix whose (i, j)-entry is 1 if $j - i = a$ and 0 otherwise. We can write $B_{ij} = \sum_{a=\lambda_j-\lambda_i}^{\lambda_j-1} b_a X^a_{\lambda_i\times\lambda_j}$ (resp. $\sum_{a=0}^{\lambda_j-1} b_a X^a_{\lambda_i\times\lambda_j}$) if $\lambda_i \leq \lambda_j$ (resp. $\lambda_i > \lambda_j$). It is easy to check that $X^a_{c\times d} X^b_{d\times e} = X^{a+b}_{c\times e}$ and this gives an algebra homomorphism from

$$\begin{pmatrix} \mathbb{F}_q[t] & t^{\lambda_2-\lambda_1}\mathbb{F}_q[t] & t^{\lambda_3-\lambda_1}\mathbb{F}_q[t] & \cdots & t^{\lambda_k-\lambda_1}\mathbb{F}_q[t] \\ \mathbb{F}_q[t] & \mathbb{F}_q[t] & t^{\lambda_3-\lambda_2}\mathbb{F}_q[t] & \cdots & t^{\lambda_k-\lambda_2}\mathbb{F}_q[t] \\ \vdots & \vdots & \vdots & \ddots & \vdots \\ \mathbb{F}_q[t] & \mathbb{F}_q[t] & \mathbb{F}_q[t] & \cdots & \mathbb{F}_q[t] \end{pmatrix}$$

to $C_M(x)$ sending the element t^a in the (i, j)-entry to $X^a_{\lambda_i \times \lambda_j}$ in the (i, j)-block and extended by linearity. From our expression for the B_{ij}, it is clear that the homomorphism is surjective and the kernel is the set of matrices described by

$$\begin{pmatrix} (t^{\lambda_1}) & (t^{\lambda_2}) & \cdots & (t^{\lambda_k}) \\ (t^{\lambda_1}) & (t^{\lambda_2}) & \cdots & (t^{\lambda_k}) \\ \vdots & \vdots & \ddots & \vdots \\ (t^{\lambda_1}) & (t^{\lambda_2}) & \cdots & (t^{\lambda_k}) \end{pmatrix},$$

where (t^{λ_i}) is the ideal of $\mathbb{F}_q[t]$ generated by t^{λ_i}. The centralizer of x in the algebra M is isomorphic to the quotient algebra

$$\mathbb{F}_q[t]_\lambda = \begin{pmatrix} \mathbb{F}_q[t]_{\lambda_1} & t^{\lambda_2-\lambda_1}\mathbb{F}_q[t]_{\lambda_2} & \cdots & t^{\lambda_k-\lambda_1}\mathbb{F}_q[t]_{\lambda_k} \\ \mathbb{F}_q[t]_{\lambda_1} & \mathbb{F}_q[t]_{\lambda_2} & \cdots & t^{\lambda_k-\lambda_2}\mathbb{F}_q[t]_{\lambda_k} \\ \vdots & \vdots & \ddots & \vdots \\ \mathbb{F}_q[t]_{\lambda_1} & \mathbb{F}_q[t]_{\lambda_2} & \cdots & \mathbb{F}_q[t]_{\lambda_k} \end{pmatrix}, \tag{2.2.3}$$

where $\mathbb{F}_q[t]_{\lambda_s} = \mathbb{F}_q[t]/(t^{\lambda_s})$ is the truncated polynomial algebra. The multiplication

$$\mathbb{F}_q[t]_{\lambda_s} \times \mathbb{F}_q[t]_{\lambda_{s'}} \to \mathbb{F}_q[t]_{\lambda_{s'}}$$

is defined by multiplying the two polynomials and removing all of the monomials of degree greater than $\lambda_{s'}$.

Searching for invertible elements of $C_M(x)$ is equivalent to searching for invertible elements of $\mathbb{F}_q[t]_\lambda$. The (i, j)-entry of an arbitrary element of $\mathbb{F}_q[t]_\lambda$ corresponds to a pair (λ_i, λ_j).

Since the λ_i are not necessarily distinct, it is convenient to redefine the notation as follows: let $\lambda_1 < \lambda_2 < \cdots < \lambda_h$ be the distinct dimensions of the Jordan blocks of x and let l_i be the multiplicity of λ_i where $1 \le i \le h$. We assemble the entries sharing the same values (λ_i, λ_j) in a unique $l_i \times l_j$ block. With this new notation, the definition of $\mathbb{F}_q[t]_\lambda$ becomes

$$\mathbb{F}_q[t]_\lambda = \begin{pmatrix} M_{l_1\times l_1}(\mathbb{F}_q[t]_{\lambda_1}) & M_{l_1\times l_2}(t^{\lambda_2-\lambda_1}\mathbb{F}_q[t]_{\lambda_2}) & \cdots & M_{l_1\times l_h}(t^{\lambda_h-\lambda_1}\mathbb{F}_q[t]_{\lambda_h}) \\ M_{l_2\times l_1}(\mathbb{F}_q[t]_{\lambda_1}) & M_{l_2\times l_2}(\mathbb{F}_q[t]_{\lambda_2}) & \cdots & M_{l_2\times l_h}(t^{\lambda_h-\lambda_2}\mathbb{F}_q[t]_{\lambda_h}) \\ \vdots & \vdots & \ddots & \vdots \\ M_{l_h\times l_1}(\mathbb{F}_q[t]_{\lambda_1}) & M_{l_h\times l_2}(\mathbb{F}_q[t]_{\lambda_2}) & \cdots & M_{l_h\times l_h}(\mathbb{F}_q[t]_{\lambda_h}) \end{pmatrix},$$

where $M_{l_i \times l_j}(F)$ is the set of $l_i \times l_j$ matrices with coefficients in F. An arbitrary element of $\mathbb{F}_q[t]_\lambda$ can be written as a block matrix

$$A = \begin{pmatrix} A_{11} & \cdots & A_{1h} \\ \vdots & \ddots & \vdots \\ A_{h1} & \cdots & A_{hh} \end{pmatrix}, \tag{2.2.4}$$

where $A_{ij} \in M_{l_i \times l_j}(t^{\lambda_j - \lambda_i}\mathbb{F}_q[t]_{\lambda_j})$ with $t^{\lambda_j - \lambda_i} = 1$ if $\lambda_i > \lambda_j$. Such A_{ij} can be written as

$$A_{ij} = A_{ij}^{(0)} + tA_{ij}^{(1)} + t^2 A_{ij}^{(2)} + t^3 A_{ij}^{(3)} + \cdots + t^{\lambda_j - 1} A_{ij}^{(\lambda_j - 1)}, \tag{2.2.5}$$

where $A_{ij}^{(s)}$ is the $l_i \times l_j$ matrix of the coefficients of t^s in the entries of A_{ij}.

Now take an arbitrary $A \in \mathbb{F}_q[t]_\lambda$ as in (2.2.4), and write it as

$$A = A^{(0)} + tA^{(1)} + \cdots + t^{\lambda_h - 1} A^{(\lambda_h - 1)}$$

with

$$A^{(s)} = \begin{pmatrix} A_{11}^{(s)} & \cdots & A_{1h}^{(s)} \\ \vdots & \ddots & \vdots \\ A_{h1}^{(s)} & \cdots & A_{hh}^{(s)} \end{pmatrix}, \tag{2.2.6}$$

where the $A_{ij}^{(s)}$ are those defined in (2.2.5). It is clear that A is invertible in $\mathbb{F}_q[t]_\lambda$ if and only if $A^{(0)}$ is invertible. In $\mathbb{F}_q[t]_\lambda$ polynomials in the blocks above the main diagonal have leading term zero—in other words,

$$A^{(0)} = \begin{pmatrix} A_{11}^{(0)} & & 0 \\ \vdots & \ddots & \\ A_{h1}^{(0)} & \cdots & A_{hh}^{(0)} \end{pmatrix}. \tag{2.2.7}$$

Thus A is invertible if and only if $A_{ii}^{(0)}$ is invertible for every i, equivalently $A_{ii}^{(0)} \in \mathrm{GL}_{l_i}(q)$.

Define the following subgroups of $\mathbb{F}_q[t]_\lambda^*$, the unit group of $\mathbb{F}_q[t]_\lambda$:

- R is the subgroup of $\mathbb{F}_q[t]_\lambda^*$ consisting of the matrices of the form

$$\begin{pmatrix} A_{11}^{(0)} & & \\ & \ddots & \\ & & A_{hh}^{(0)} \end{pmatrix}$$

  with $A_{ii}^{(0)} \in \mathrm{GL}_{l_i}(q)$.

- U is the subgroup of $\mathbb{F}_q[t]^*_\lambda$ consisting of the matrices of the form (2.2.4) where $A_{ii}^{(0)}$ is the identity matrix for every i; equivalently, U consists of the matrices

$$\begin{pmatrix} 1+tU_{11} & U_{12} & \cdots & U_{1h} \\ U_{21} & \ddots & \ddots & \vdots \\ \vdots & \ddots & \ddots & U_{h-1,h} \\ U_{h1} & \cdots & U_{h,h-1} & 1+tU_{hh} \end{pmatrix},$$

where U_{ij} is an arbitrary $l_i \times l_j$ block with entries in $t^{\lambda_j-\lambda_i}\mathbb{F}_q[t]_{\lambda_j}$ (with $t^{\lambda_j-\lambda_i} = 1$ if $\lambda_i > \lambda_j$).

Proposition 2.2.2 *Let R and U be defined as above.*

(i) $R \cap U$ is the trivial subgroup.
(ii) $\mathbb{F}_q[t]^*_\lambda \subseteq UR$.
(iii) U is a normal subgroup of UR.

Hence $\mathbb{F}_q[t]^*_\lambda = U \rtimes R$.

Proof Part (i) is clear. For (ii), let $A \in \mathbb{F}_q[t]^*_\lambda$ and write $A = A^{(0)} + tA^{(1)} + t^2A^{(2)} + \cdots$ as in (2.2.6). If D is the block diagonal matrix $\bigoplus_{i=1}^h A_{ii}^{(0)}$, where $A_{ii}^{(0)}$ are as in (2.2.7), then $D \in R$ and $AD^{-1} \in U$.

Finally, consider (iii). Let $B \in U$ and $X \in \mathbb{F}_q[t]^*_\lambda$ arbitrary. Write $B = B^{(0)} + tB^{(1)} + t^2B^{(2)} + \cdots$ and $X = X^{(0)} + tX^{(1)} + t^2X^{(2)} + \cdots$ as we did for A. Note that $X^{-1} = {X^{(0)}}^{-1} + tX'$ for some $X' \in \mathbb{F}_q[t]_\lambda$, and $X^{-1}BX = {X^{(0)}}^{-1}B^{(0)}X^{(0)} + tB'$ for some $B' \in \mathbb{F}_q[t]_\lambda$. As shown in (2.2.7), $X^{(0)}$ and $B^{(0)}$ are lower triangular block matrices; so is ${X^{(0)}}^{-1}B^{(0)}X^{(0)}$. Moreover, the blocks on the main diagonal of ${X^{(0)}}^{-1}B^{(0)}X^{(0)}$ are ${X_{ii}^{(0)}}^{-1}B_{ii}^{(0)}X_{ii}^{(0)} = 1$, since $B_{ii}^{(0)} = 1$. This is exactly the condition $X^{-1}BX \in U$. □

As in the previous chapter, in the next result we use the notation $\bigoplus_{i=1}^h J_{\lambda_i}^{[l_i]}$ to denote a block diagonal matrix having l_i diagonal blocks J_{λ_i} for $i = 1\ldots, h$.

Theorem 2.2.3 *Let $x = \bigoplus_{i=1}^h J_{\lambda_i}^{[l_i]}$ be a unipotent element of $\mathrm{GL}_n(q)$. Then $C_{\mathrm{GL}_n(q)}(x) = U \rtimes R$, where $R \cong \prod_{i=1}^h \mathrm{GL}_{l_i}(q)$ and $|U| = q^m$ with*

$$m = 2\sum_{i<j} \lambda_i l_i l_j + \sum_i (\lambda_i - 1)l_i^2. \tag{2.2.8}$$

Proof Let $G = \mathrm{GL}_n(q)$. Working in $\mathbb{F}_q[t]_\lambda$ instead of $C_G(x)$, the groups R and U are the subgroups described above, and $C_G(x) \cong U \rtimes R$. That $R \cong \prod_{i=1}^h \mathrm{GL}_{l_i}(q)$ follows by definition of R.

It remains to compute the cardinality of U, for which we need to determine how many choices there are for the U_{ij}. The term $1 + tU_{ii}$ equals $1 + tA_{ii}^{(1)} + \cdots + t^{\lambda_i - 1}A_{ii}^{(\lambda_i - 1)}$ and $A_{ii}^{(s)}$ is a matrix in $M_{l_i \times l_i}(\mathbb{F}_q)$ which can be chosen arbitrarily; thus, for U_{ii} there are $q^{(\lambda_i - 1)l_i^2}$ choices and these give the second sum in (2.2.8).

If $i > j$, then $\lambda_i > \lambda_j$ and $U_{ij} = A_{ij}^{(0)} + \cdots + t^{\lambda_j - 1}A_{ij}^{(\lambda_j - 1)}$. Every matrix $A_{ij}^{(s)}$ is an arbitrary $l_i \times l_j$ matrix with entries in $\mathbb{F}_q$, so the number of possible choices for U_{ij} is $q^{\lambda_i l_i l_j}$.

If $i < j$, then $\lambda_i < \lambda_j$ and $U_{ij} = t^{\lambda_j - \lambda_i}A_{ij}^{(\lambda_j - \lambda_i)} + \cdots + t^{\lambda_j - 1}A_{ij}^{(\lambda_j - 1)}$. Again, $A_{ij}^{(s)}$ is an arbitrary $l_i \times l_j$ matrix with entries in $\mathbb{F}_q$, so the number of possible choices for U_{ij} is $q^{\lambda_j l_i l_j}$.

Summing over all $i \neq j$, we get the first sum in (2.2.8). □

2.2.3 Generators for the Centralizer of a Unipotent Element

As above, let $x = \bigoplus_{i=1}^{h} J_{\lambda_i}^{[l_i]}$ be a unipotent element of $\mathrm{GL}_n(q)$. We now describe a generating set for $C_{\mathrm{GL}_n(q)}(x)$ by working in $\mathbb{F}_q[t]_\lambda$. We continue the notation of the previous section.

Observe that every element of $\mathbb{F}_q[t]_\lambda$ can be written as a block matrix $A = (A_{ij})$ as in (2.2.4). For example, if A is the identity matrix, then A_{ii} is the $l_i \times l_i$ identity matrix with coefficients in $\mathbb{F}_q[t]_{\lambda_i}$ and $A_{ij} = 0$ for $i \neq j$.

We introduce the following notation:

- $\mathsf{a}_i(y)$, for $1 \leq i \leq h$ and $y \in \mathrm{GL}_{l_i}(\mathbb{F}_q[t]_{\lambda_i})$, is the matrix obtained by taking the identity matrix in $\mathbb{F}_q[t]_\lambda$ and replacing the block A_{ii} by y;
- $\mathsf{b}_i(\mu)$, for $1 \leq i \leq h$ and $\mu \in \mathbb{F}_q[t]_{\lambda_i}$, is the matrix obtained by taking the identity matrix in $\mathbb{F}_q[t]_\lambda$ and replacing the block A_{ii} by the diagonal matrix

$$\begin{pmatrix} \mu & & & \\ & 1 & & \\ & & \ddots & \\ & & & 1 \end{pmatrix};$$

- $\mathsf{c}_i^+(\mu)$ (resp. $\mathsf{c}_i^-(\mu)$), for $1 \leq i \leq h-1$ and $\mu \in \mathbb{F}_q[t]_{\lambda_{i+1}}$ (resp. $\mathbb{F}_q[t]_{\lambda_i}$), is the matrix obtained by taking the identity matrix in $\mathbb{F}_q[t]_\lambda$ and replacing the block $A_{i,i+1}$ (resp. $A_{i+1,i}$) by the matrix having μ in the bottom-left (resp. top-right) corner and 0 elsewhere. We write c_i^+ and c_i^- for $\mathsf{c}_i^+(1)$ and $\mathsf{c}_i^-(1)$ respectively.

Recall, for example from [17, p. 59], that a *primitive element* of $\mathbb{F}_q$ is a generator for the cyclic group $\mathbb{F}_q^*$.

Theorem 2.2.4 *Let ω be a primitive element of $\mathbb{F}_q$. The following collection of matrices forms a generating set for the unit group $\mathbb{F}_q[t]^*_\lambda$:*

- *the set of $\mathsf{a}_i(y)$, where $1 \leq i \leq h$ and y runs over a generating set for $\mathrm{GL}_{l_i}(q)$ (as subgroup of $\mathrm{GL}_{l_i}(\mathbb{F}_q[t]_{\lambda_i})$;*
- *the set of $\mathsf{b}_i(\mu)$, where $1 \leq i \leq h$ and μ runs over all elements of the set*

$$\{1 + \omega^j t^l : 0 \leq j \leq [\mathbb{F}_q : \mathbb{F}_p], 1 \leq l \leq \lambda_i - 1\};$$

- *the set of all c_i^+ and c_i^-, where $1 \leq i \leq h - 1$.*

To prove the theorem we first establish some preliminary facts. The following lemma shows that, with the generators listed above, we can get all elements of the form $\mathsf{a}_i(y)$ for $y \in \mathrm{GL}_{l_i}(q)$, and all elements of the form $\mathsf{b}_i(\mu)$ for $\mu \in \mathbb{F}_q[t]_{\lambda_i}$.

Lemma 2.2.5 *Let ω be a primitive element of $\mathbb{F}_q$. A generating set for $\mathbb{F}_q[t]^*_{\lambda_i}$ is given by*

$$\Delta := \{\omega\} \cup \{1 + \omega^j t^l : 0 \leq j \leq [\mathbb{F}_q : \mathbb{F}_p], 1 \leq l \leq \lambda_i - 1\}. \tag{2.2.9}$$

Proof Let $r = [\mathbb{F}_q : \mathbb{F}_p]$. Let $a_0 + a_1 t + a_2 t^2 + \cdots + a_{\lambda_i} t^{\lambda_i}$ be an element of $\mathbb{F}_q[t]^*_{\lambda_i}$, with $a_j \in \mathbb{F}_q$ and $a_0 \neq 0$. We claim that for any $m \leq \lambda_i$ there is a product of elements of Δ such that the first $m + 1$ coefficients coincide with $a_0, \ldots, a_m$. We prove this by induction on m.

For $m = 0$, this is obvious since $a_0 \neq 0$, so a_0 is a power of ω. Now suppose there exists a product $P(t)$ of elements of Δ, say $P(t) = a_0 + a_1 t + \cdots + a_m t^m + b_{m+1} t^{m+1} + \cdots + b_{\lambda_i} t^{\lambda_i}$ for some $b_{m+1}, \ldots, b_{\lambda_i} \in \mathbb{F}_q$. Since $a_{m+1} - b_{m+1} \in \mathbb{F}_q$, we can find $y_0, \ldots, y_{r-1} \in \mathbb{F}_p$ such that $a_0^{-1}(a_{m+1} - b_{m+1}) = y_0 + y_1\omega + y_2\omega^2 + \cdots + y_{r-1}\omega^{r-1}$. A straightforward computation shows that

$$\begin{aligned}
&P(t) \cdot \prod_{j=0}^{r-1} (1 + \omega^j t^{m+1})^{y_j} \\
&= P(t) \cdot (1 + t^{m+1}(y_0 + y_1\omega + \cdots + y_{r-1}\omega^{r-1}) + t^{m+2} Q(t)) \\
&= (a_0 + a_1 t + \cdots + a_m t^m + b_{m+1} t^{m+1} + t^{m+2} R(t)) \cdot \\
&\quad (1 + a_0^{-1}(a_{m+1} - b_{m+1}) t^{m+1} + t^{m+2} Q(t)) \\
&= a_0 + a_1 t + \cdots + a_{m+1} t^{m+1} + t^{m+2} S(t),
\end{aligned}$$

where $Q(t)$, $R(t)$ and $S(t)$ are elements of $\mathbb{F}_q[t]$. Hence, we have exhibited a product of elements of Δ whose first $m + 1$ coefficients are $a_0, \ldots, a_{m+1}$. □

Proposition 2.2.6 *Every diagonal matrix in $\mathbb{F}_q[t]_\lambda^*$ can be written as a product of elements of the form $\mathsf{a}_i(y)$ for $y \in \mathrm{GL}_{l_i}(q)$ and $\mathsf{b}_i(\mu)$ for $\mu \in \mathbb{F}_q[t]_{\lambda_i}$.*

Proof It is sufficient to prove that we can build every diagonal matrix where all but one of the elements on the main diagonal are 1. Say D is a diagonal matrix whose diagonal has $\mu \neq 0, 1$ in a certain entry and 1 elsewhere. If μ belongs to the block A_{ii}, then D can be obtained by conjugating the matrix $\mathsf{b}_i(\mu)$ by an appropriate permutation matrix in the block A_{ii} (permutation matrices have coefficients in $\mathbb{F}_q$, so they are products of elements of the form $\mathsf{a}_i(y)$). □

For the rest of the section, we regard elements of $\mathbb{F}_q[t]_\lambda$ as $k \times k$ matrices, as in (2.2.3). Let $\mathbb{I}$ be the identity matrix in $\mathbb{F}_q[t]_\lambda$ and let $E_{j\ell}$ be the matrix in $\mathbb{F}_q[t]_\lambda$ having 1 in the (j, ℓ)-entry where $1 \leq j, \ell \leq k$ and 0 elsewhere.

Proposition 2.2.7 *Every matrix of the form $\mathbb{I} + \nu E_{j\ell}$, with $1 \leq j, \ell \leq k$ and $\nu \in \mathbb{F}_q[t]_{\lambda_s}$ for some s, can be obtained as a product of elements of the form $\mathsf{a}_i(y)$, $\mathsf{b}_i(\mu)$, c_i^+ and c_i^-.*

Proof Consider the elements with $j < \ell$ (namely, the entry ν is above the main diagonal). Observe first that every matrix of the form $\mathbb{I} + E_{j,j+1}$ is either an element of the form $\mathsf{a}_i(y)$ (if the $(j, j+1)$-entry is in one of the blocks A_{ii}) or c_i^+ (otherwise). If $j < m < \ell$, then computation shows that

$$(\mathbb{I} + E_{jm})(\mathbb{I} + E_{m\ell})(\mathbb{I} - E_{jm})(\mathbb{I} - E_{m\ell}) = \mathbb{I} + E_{j\ell};$$

this shows by induction on $|\ell - j|$ that every element of the form $\mathbb{I} + E_{j\ell}$ can be obtained as a product of the given generators. Finally, for every invertible $\nu \in \mathbb{F}_q[t]_{\lambda_s}$ and $1 \leq j, \ell \leq k$, observe that

$$\mathbb{I} + \nu E_{j\ell} = (\mathbb{I} + (\nu - 1)E_{jj})(\mathbb{I} + E_{j\ell})(\mathbb{I} + (\nu^{-1} - 1)E_{jj}),$$

where the second factor is obtained as above, and the first and third factors are obtained as explained in Proposition 2.2.6. If ν is not invertible, then it is the sum of two invertible elements ν_1, ν_2 (for example 1 and $\nu - 1$), so

$$\mathbb{I} + \nu E_{j\ell} = (\mathbb{I} + \nu_1 E_{j\ell})(\mathbb{I} + \nu_2 E_{j\ell}),$$

and both terms in the product can be obtained as above.

The argument is the same for the elements with $j > \ell$. □

Proof of Theorem 2.2.4 Let $Y = (y_{j\ell})$ be an element of the unit group $\mathbb{F}_q[t]_\lambda^*$ (so Y is $k \times k$ as in (2.2.3)). We can suppose without loss of generality that y_{11} is invertible, otherwise we switch columns by multiplying on the right by permutation matrices on the block A_{11}. Observe that

$$\prod_{j=2}^{k}(\mathbb{I}-(y_{j1}y_{11}^{-1})E_{j1})\cdot Y\cdot\prod_{j=2}^{k}(\mathbb{I}-(y_{1j}y_{11}^{-1})E_{1j})=\left(\begin{array}{c|c} y_{11} & \cdots\ 0\ \cdots \\ \hline \vdots & \ddots \\ 0 & Y' \\ \vdots & \ddots \end{array}\right).$$

We can now apply the same argument to the matrix Y' and proceed inductively until we get a diagonal matrix. Thus Y is a product of diagonal matrices, matrices of the form $\mathsf{a}_i(y)$ where y is a permutation matrix in $GL_{l_i}(q)$, and matrices of the form $\mathbb{I}+\mu E_{j\ell}$ where $\mu\in\mathbb{F}_q[t]_{\lambda_s}$ for some s. As shown in the previous results, each of these matrices can be written as a product of the generators listed in the theorem. □

2.2.4 Centralizer of a General Element

Using Theorem 2.2.3 together with (2.2.1), we deduce the following result giving the structure of the centralizer of an arbitrary element of $GL_n(q)$.

Theorem 2.2.8 *Let $x\in GL_n(q)$ have minimal polynomial $\prod_{i=1}^{h}f_i(t)^{e_i}$, where $f_1,\ldots,f_h\in\mathbb{F}_q[t]$ are distinct and irreducible, and $\deg f_i=d_i$. Let the Jordan form of x be*

$$\bigoplus_{i=1}^{h}(J_{f_i,\lambda_{i1}}^{[l_{i1}]}\oplus\cdots\oplus J_{f_i,\lambda_{ik_i}}^{[l_{ik_i}]}),$$

where $\lambda_{i1}<\cdots<\lambda_{ik_i}=e_i$ for each i. Then $C_G(x)=U\rtimes R$, where $R\cong\prod_{i=1}^{h}\left(\prod_{j=1}^{k_i}GL_{l_{ij}}(q^{d_i})\right)$ and $|U|=q^m$ with

$$m=\sum_{i=1}^{h}d_i\left(2\sum_{a<b}\lambda_{ia}l_{ia}l_{ib}+\sum_{j}(\lambda_{ij}-1)l_{ij}^2\right).$$

2.3 Conjugacy and Centralizers in $SL_n(q)$

For completeness we include the solutions of the main conjugacy problems (1)–(3) of Sect. 1.1 for the special linear group. In particular, we construct a generating set for the centralizer in $SL_n(q)$ of an arbitrary element, since this is not immediate from our solution for $GL_n(q)$.

Theorem 2.3.1 *Let $x \in \mathrm{SL}_n(q)$ have minimal polynomial $\prod_{i=1}^h f_i(t)^{e_i}$, where $f_1, \dots, f_h \in \mathbb{F}_q[t]$ are distinct and irreducible, and $\deg f_i = d_i$. Let the Jordan form of x be*

$$\bigoplus_{i=1}^{h}(J_{f_i,\lambda_{i1}}^{[l_{i1}]} \oplus \cdots \oplus J_{f_i,\lambda_{ik_i}}^{[l_{ik_i}]}), \tag{2.3.1}$$

where $\lambda_{i1} < \cdots < \lambda_{ik_i} = e_i$ for each i. Set $r = \gcd(\lambda_{11}, \dots, \lambda_{hk_h}, q-1)$. Then

$$|C_{\mathrm{GL}_n(q)}(x) : C_{\mathrm{SL}_n(q)}(x)| = \frac{q-1}{r}.$$

The conjugacy class $x^{\mathrm{GL}_n(q)}$ splits into r classes in $\mathrm{SL}_n(q)$, with representatives $x, x^z, \dots, x^{z^{r-1}}$, where $z \in \mathrm{GL}_n(q)$ is a fixed element of determinant ω (a primitive element of $\mathbb{F}_q$).

Proof We first deal with the case where $h = 1$. To simplify notation, write $f = f_1$ of degree d, and let the distinct elementary divisors of x be $f^{\lambda_1}, \dots, f^{\lambda_k}$, with multiplicities $l_1, \dots, l_k$, so that $\sum_i \lambda_i l_i = n/d$. If $x = su = us$ is the Jordan decomposition of x with s semisimple and u unipotent, then by Proposition 2.2.1, $C_{\mathrm{GL}_n(q)}(s) = \mathrm{GL}_e(q^d)$, where $de = n$. Then $u \in \mathrm{GL}_e(q^d)$ is unipotent with Jordan form $\bigoplus_i J_{\lambda_i}^{[l_i]}$. By Theorem 2.2.3,

$$C_{\mathrm{GL}_n(q)}(x) = C_{\mathrm{GL}_e(q^d)}(u) = UR,$$

where $R \cong \prod_{i=1}^k \mathrm{GL}_{l_i}(q^d)$. If we view V as $V_e(q^d)$, then the action of R on V is as

$$\prod_{i=1}^{k}(\mathrm{GL}_{l_i}(q^d) \otimes I_{\lambda_i}).$$

Let ν be a primitive element of $\mathbb{F}_{q^d}$ such that $\mathrm{N}(\nu) = \omega$, where N is the norm map $\mathbb{F}_{q^d} \to \mathbb{F}_q$. If $A \in \mathrm{GL}_{l_i}(q^d)$ has determinant ν, then the determinant of $A \otimes I_{\lambda_i}$ as an element of $\mathrm{GL}_e(q^d)$ is ν^{λ_i}, and so its determinant as an element of $\mathrm{GL}_n(q)$ is $\mathrm{N}(\nu^{\lambda_i})$ (see [21, (4.3.13)]), which is ω^{λ_i}.

Thus the subgroup of $\mathbb{F}_q^*$ consisting of all determinants of elements of R is generated by ω^{λ_i} for $i = 1, \dots, k$. Hence it equals $\langle \omega^r \rangle$ where $r = \gcd(\lambda_1, \dots, \lambda_k, q-1)$. Thus the image of the determinant map $C_{\mathrm{GL}_n(q)}(x) \to \mathbb{F}_q^*$ has order $(q-1)/r$. This completes the proof when $h = 1$.

Now consider the general case. We know from (2.2.1) that $C_{\mathrm{GL}(V)}(x) = \prod_{i=1}^h C_{\mathrm{GL}(V_i)}(x_i)$, where $V_i = \ker f_i(x)^{e_i}$ are the generalized eigenspaces of x. By the $h = 1$ case, the image of the determinant map on $C_{\mathrm{GL}(V_i)}(x_i)$ is $\langle \omega^{r_i} \rangle$, where

$r_i = \gcd(\lambda_{i1}, \ldots, \lambda_{ik_i}, q-1)$. It follows that the image of the determinant map on $C_{\mathrm{GL}(V)}(x)$ is $\langle \omega^r \rangle$, where $r = \gcd(r_1, \ldots, r_h)$. The conclusion follows. □

We now use this theorem to decide conjugacy and compute conjugating elements in $S := \mathrm{SL}_n(q)$. First, as conjugacy class representatives in S, for each Jordan form $J \in S$ as in (2.3.1), we choose representatives

$$J, J^z, \ldots, J^{z^{r-1}},$$

where $r = \gcd(\lambda_{11}, \ldots, \lambda_{hk_h}, q-1)$ and $z \in \mathrm{GL}_n(q)$ is a fixed element of determinant ω (a primitive element of $\mathbb{F}_q$).

Given $x \in S$ with Jordan form J as in (2.3.1), we find its class representative by computing $g \in \mathrm{GL}_n(q)$ such that $J = x^g$; then x is S-conjugate to J^{z^i}, where $\det(g) \in \omega^{-i}\langle \omega^r \rangle$.

To decide whether $x, y \in S$, both with Jordan form (2.3.1), are S-conjugate, we compute $g \in \mathrm{GL}_n(q)$ such that $y = x^g$; then x and y are S-conjugate if and only if $\det(g) \in \langle \omega^r \rangle$.

Finally, given $x \in S$ that is S-conjugate to a representative J^{z^i}, we find a conjugating element as follows. First, compute $g \in \mathrm{GL}_n(q)$ such that $x^g = J$. Then $\det(g) \in \omega^{-i}\langle \omega^r \rangle$, say $\det(g) = \omega^{-i+kr}$. From the proof of Theorem 2.3.1, we can find $c \in C_{\mathrm{GL}_n(q)}(J)$ of determinant ω^{-kr} with c a product of scalars in the factors $\mathrm{GL}_{l_{ij}}(q^{d_i})$ of the subgroup R of $C(J)$. Then $x^{gcz^i} = J^{z^i}$, and $gcz^i \in S$.

2.3.1 Generators for Centralizer in $\mathrm{SL}_n(q)$

Write $G = \mathrm{GL}_n(q)$ and $S = \mathrm{SL}_n(q)$, and let $x \in S$ be as in (2.3.1). Using the notation of Theorem 2.2.8, $C_G(x) = U \rtimes R$, and so $C_S(x) = U \rtimes (R \cap S)$, since every element of U has determinant 1. So our task is to compute the elements of R of determinant 1.

Theorem 2.2.8 shows that there is an isomorphism ϕ between R and $\prod_{i=1}^h \mathrm{GL}_{l_{ij}}(q^{d_i})$. We realise this isomorphism as follows. Map a sequence $[z_{ij} : 1 \le i \le h, 1 \le j \le k_i]$ with $z_{ij} \in \mathrm{GL}_{l_{ij}}(q^{d_i})$ to the block diagonal matrix

$$\begin{pmatrix} \ddots & & \\ & Z_{ij} & \\ & & \ddots \end{pmatrix}, \tag{2.3.2}$$

where Z_{ij} is defined as follows: if $z_{ij} = (\alpha_{\mu\nu})$, then Z_{ij} is the matrix obtained by substituting each entry $\alpha_{\mu\nu}$ of z_{ij} by the embedding into $\mathrm{GL}_{\lambda_{ij}d_i}(q)$ of a λ_{ij}-dimensional scalar matrix with diagonal $\alpha_{\mu\nu}$. Note that if $z \in R$ is such that

$z = \varphi(z_{11}, \ldots, z_{hk_h})$, where $z_{ij} \in \mathrm{GL}_{l_{ij}}(q^{d_i})$ corresponds to a Jordan block of dimension λ_{ij}, then

$$\det(z) = \prod_{i,j} \mathrm{N}_i(\det(z_{ij}))^{\lambda_{ij}}, \tag{2.3.3}$$

where $\mathrm{N}_i : \mathbb{F}_{q^{d_i}} \to \mathbb{F}_q$ is the norm map. Let ω be a primitive element of $\mathbb{F}_q$. Then $\mathrm{N}_i(\det(z_{ij})) = \omega^{a_{ij}}$ for some $a_{ij} \in \mathbb{Z}_{q-1}$, the ring of residue classes modulo $q-1$, so (2.3.3) becomes $\det(z) = \prod_{i,j} \omega^{a_{ij}\lambda_{ij}}$. The condition $\det(z) = 1$ is equivalent to $\sum_{i=1}^{h} \sum_{j=1}^{k_i} a_{ij}\lambda_{ij} = 0$.

Let V_i be the generalized eigenspace $\ker f_i(x)^{e_i}$, so that $x = x_1 \oplus \cdots \oplus x_h$, where x_i is the restriction of x to V_i. Every $y \in C_S(x)$ can also be written as $y = y_1 \oplus \cdots \oplus y_h$, but the centralizer in S of x is not necessarily the direct product of the centralizers of the x_i. We list generators for $C_S(x)$ in two steps.

1. For each i, we generate $C_{S_i}(x_i)$, where $S_i = \mathrm{SL}(V_i) \cong \mathrm{SL}_{n_i d_i}(q)$ and $n_i = \sum_j l_{ij}\lambda_{ij}$. Putting them together, we generate a subgroup

$$\prod_{i=1}^{h} C_{S_i}(x_i) \tag{2.3.4}$$

   of $C_S(x)$.
2. To complete a generating set for $C_S(x)$, we add elements

$$\mathrm{h}_m = \bigoplus_{i,j} \mathrm{h}_{mij}, \ 1 \le m \le s,$$

   where each h_{mij} commutes with $J_{f_i,\lambda_{ij}}^{[l_{ij}]}$, and together with the subgroup (2.3.4), $\mathrm{h}_1, \ldots, \mathrm{h}_s$ generate the kernel of the determinant map $R \to \mathbb{F}_q^*$.

We now describe how to write down generators for each of the above steps: call $\mathcal{S}_i$ the set of generators in Step i for $i = 1, 2$.

We first discuss how to construct $\mathcal{S}_2$. For every $1 \le i \le h$ and $1 \le j \le k_i$, take $\widetilde{\mathrm{h}}_{ij} \in \mathrm{GL}_{l_{ij}}(q^{d_i})$ having determinant of order $q^{d_i} - 1$ and such that its image h_{ij} under application of the homomorphism φ defined in (2.3.2) has determinant $\omega^{\lambda_{ij}}$. Let $k = \sum_{i=1}^{h} k_i$ and let $Z(0)$ be the subgroup of $\mathbb{Z}_{q-1}^k$ consisting of the solutions $(x_{11}, \ldots, x_{hk_h})$ of the equation $\sum_{i,j} x_{ij}\lambda_{ij} = 0$ in $\mathbb{Z}_{q-1}$. Compute a generating set $\{(a_{m11}, \ldots, a_{mhk_h}) : 1 \le m \le s\}$ for $Z(0)$. Then $\mathcal{S}_2$ consists of all the matrices

$$\mathrm{h}_m = \bigoplus_{i,j} \mathrm{h}_{ij}^{a_{mij}}, \quad 1 \le m \le s.$$

We construct $\mathcal{S}_1$ by describing separately each $C_{S_i}(x_i)$ for every i. Let $d_i = \deg f_i$ and let $n_i = \sum_{j=1}^{k_i} \lambda_{ij} l_{ij}$. Let x_i be the embedding into $\mathrm{GL}_{n_i d_i}(q)$ of some $\widetilde{x}_i \in \mathrm{GL}_{n_i}(q^{d_i})$ and let $C_{\mathrm{GL}_{n_i d_i}(q)}(x_i)$ be the embedding into $\mathrm{GL}_{n_i d_i}(q)$ of $C_{\mathrm{GL}_{n_i}(q^{d_i})}(\widetilde{x}_i)$. If $\widetilde{x}_i = \widetilde{s}_i \widetilde{u}_i$ is the Jordan decomposition of $\widetilde{x}_i$, then $\widetilde{s}_i$ is a scalar matrix, so $C_{\mathrm{GL}_{n_i}(q^{d_i})}(\widetilde{x}_i) = C_{\mathrm{GL}_{n_i}(q^{d_i})}(\widetilde{u}_i)$. So we reduce our task to computing the centralizer of a unipotent element with Jordan form $\bigoplus_{j=1}^{k_i} J_{\lambda_{ij}}^{[l_{ij}]}$. For every $y \in C_{\mathrm{GL}_{n_i d_i}(q)}(x_i)$, if $\widetilde{y}$ is the corresponding element in $C_{\mathrm{GL}_{n_i}(q^{d_i})}(\widetilde{x}_i)$, then $\det(y) = \mathrm{N}_i(\det(\widetilde{y}))$. Hence $\det(y) = 1$ if and only if $\det(\widetilde{y})$ has determinant a power of ω_i^{q-1}, where ω_i is a primitive element of $\mathbb{F}_{q^{d_i}}$. It follows that the construction of a generating set for $C_{S_i}(x_i)$ reduces to listing the generators of the centralizer of a unipotent element in a subgroup of a specific index in $\mathrm{GL}_{n_i}(q^{d_i})$. We do this by modifying Theorem 2.2.4.

To state this modification, we revert to the notation of Theorem 2.2.4. Thus let $x = \bigoplus_{i=1}^{h} J_{\lambda_i}^{[l_i]}$ be a unipotent element of $\mathrm{GL}_n(q)$ with corresponding partition λ of n, and recall that we have an isomorphism between the centralizer $C_{\mathrm{GL}_n(q)}(x)$ and the unit group $\mathbb{F}_q[t]_\lambda^*$.

Theorem 2.3.2 *Let ω be a primitive element of $\mathbb{F}_q$ and let d be a divisor of $q-1$. A generating set for the subgroup of $\mathbb{F}_q[t]_\lambda^*$ of matrices whose determinant is a power of ω^d consists of the following:*

(1) *the set of $\mathsf{a}_i(y)$, where $1 \le i \le h$ and y runs over a generating set for the unique subgroup of $\mathrm{GL}_{l_i}(q)$ of index d;*
(2) *the set of $\mathsf{b}_i(\mu)$, where $1 \le i \le h$ and μ runs over all elements of the set*

$$\{1 + \omega^j t^l : 0 \le j \le [\mathbb{F}_q : \mathbb{F}_p], 1 \le l \le \lambda_i - 1\};$$

(3) *the set of all $\mathsf{c}_i^+(\xi)$ and $\mathsf{c}_i^-(\xi)$, where $1 \le i \le h-1$ and $\xi \in \{1, \omega, \ldots, \omega^{[\mathbb{F}_q:\mathbb{F}_p]-1}\}$;*
(4) *matrices which, together with the elements in (1) – (3) above, generate the kernel of the determinant map $R \to \mathbb{F}_q^*$.*

Proof The proof is similar to that of Theorem 2.2.4. Using elements of the form $\mathsf{a}_i(y)$ and $\mathsf{b}_i(\mu)$ described above, we can get in every block A_{ii} a matrix whose determinant is a power of ω^d. To get diagonal elements, here we cannot conjugate by permutation matrices (their determinant is not a power of ω^d in general), but we can conjugate by a monomial matrix of determinant 1.

The elements c_i^+ and c_i^- are not sufficient to repeat the argument of Theorem 2.2.4, because we do not have diagonal matrices to get all elements of the form $\mathbb{I} + \nu E_{j\ell}$. But for every $1 \le j, \ell \le h$,

$$(\mathbb{I} + \xi_1 E_{j\ell})(\mathbb{I} + \xi_2 E_{j\ell}) = \mathbb{I} + (\xi_1 + \xi_2) E_{j\ell}$$

for every $\xi_1 \in \mathbb{F}_q[t]_{\lambda_{s_1}}$ and $\xi_2 \in \mathbb{F}_q[t]_{\lambda_{s_2}}$ for appropriate s_1 and s_2. So we take $\mathsf{c}_i^+(\xi)$ and $\mathsf{c}_i^-(\xi)$ for enough ξ to generate the additive group of $\mathbb{F}_q$.

Finally, using the matrices listed at (1)–(3) we can generate every matrix where each block A_{ii} has determinant a power of ω^d. But we need additional matrices satisfying the weaker condition that $\prod_i \det(A_{ii})$ is a power of ω^d. For these we add elements listed at (4); they are constructed using the method described for $\mathcal{S}_2$. □

2.4 Some Examples

We illustrate some of the results of this chapter with the following examples. Matrices are listed in Jordan form and we use the notation of the earlier sections throughout.

Example 2.4.1 We list the conjugacy classes of $\mathrm{GL}_3(2)$. If $x \in \mathrm{GL}_3(2)$, then its characteristic polynomial has degree 3 and it is a product of irreducible polynomials of degree at most 3 in $\mathbb{F}_2[t]$. These are $t+1$, t^2+t+1, t^3+t+1 and t^3+t^2+1. Every elementary divisor is a power of one of these polynomials and the sum of the degrees must be 3. In Table 2.1, for each class representative, we record its elementary divisors, its Jordan form, and its minimal and characteristic polynomials.

Table 2.1 Conjugacy classes in $\mathrm{GL}_3(2)$

List of elementary divisors	Representative	Minimal polynomial	Characteristic polynomial
$t+1$, $t+1$, $t+1$	$\begin{pmatrix} 1&0&0\\0&1&0\\0&0&1 \end{pmatrix}$	$t+1$	$(t+1)^3$
$t+1$, $(t+1)^2$	$\begin{pmatrix} 1&0&0\\0&1&1\\0&0&1 \end{pmatrix}$	$(t+1)^2$	$(t+1)^3$
$(t+1)^3$	$\begin{pmatrix} 1&1&0\\0&1&1\\0&0&1 \end{pmatrix}$	$(t+1)^3$	$(t+1)^3$
$t+1$, t^2+t+1	$\begin{pmatrix} 1&0&0\\0&0&1\\0&1&1 \end{pmatrix}$	$(t+1)(t^2+t+1)$	$(t+1)(t^2+t+1)$
t^3+t+1	$\begin{pmatrix} 0&1&0\\0&0&1\\1&1&0 \end{pmatrix}$	t^3+t+1	t^3+t+1
t^3+t^2+1	$\begin{pmatrix} 0&1&0\\0&0&1\\1&0&1 \end{pmatrix}$	t^3+t^2+1	t^3+t^2+1

Example 2.4.2 Consider $x \in G := \mathrm{GL}_{12}(3)$ where

$$x = \begin{pmatrix}
2&0&0&0&0&0&0&0&0&0&0&0\\
0&0&1&0&0&0&0&0&0&0&0&0\\
0&1&1&0&0&0&0&0&0&0&0&0\\
0&0&0&0&1&0&0&0&0&0&0&0\\
0&0&0&1&1&0&0&0&0&0&0&0\\
0&0&0&0&0&0&1&1&0&0&0&0\\
0&0&0&0&0&1&1&0&1&0&0&0\\
0&0&0&0&0&0&0&0&1&0&0&0\\
0&0&0&0&0&0&0&1&1&0&0&0\\
0&0&0&0&0&0&0&0&0&0&1&0\\
0&0&0&0&0&0&0&0&0&0&0&1\\
0&0&0&0&0&0&0&0&0&1&2&2
\end{pmatrix}.$$

The minimal polynomial of x is $(t-2)(t^2-t-1)^2(t^3+t^2+t-1)$. The generalized eigenspaces are $V_1 = \ker(x-2)$, $V_2 = \ker((x^2-x-1)^2)$ and $V_3 = \ker(x^3+x^2+x-1)$. Recall that $C_G(x) = U \rtimes R$.

Since V_1 has dimension 1, the corresponding factor in R is $\mathrm{GL}_1(3)$.

The matrix of the restriction of x to V_2 is the embedding in $\mathrm{GL}_8(3)$ of

$$y := \begin{pmatrix}
\omega&0&0&0\\
0&\omega&0&0\\
0&0&\omega&1\\
0&0&0&\omega
\end{pmatrix} \in \mathrm{GL}_4(9),$$

where ω is a root of the polynomial t^2-t-1 in $\mathbb{F}_9$. Observe y has two Jordan blocks of order 1 and one block of order 2. An element in $\mathrm{GL}_4(9)$ centralizing y has the form

$$\left(\begin{array}{cc|cc}
\multicolumn{2}{c|}{\multirow{2}{*}{A}} & 0 & * \\
 & & 0 & * \\
\hline
* & * & \multicolumn{2}{c}{\multirow{2}{*}{X}} \\
0 & 0 & &
\end{array}\right),$$

where $A \in \mathrm{GL}_2(9)$ and X is a matrix of the form $b_0 X^0_{2\times 2} + b_1 X^1_{2\times 2}$ with $b_0, b_1 \in \mathbb{F}_9$ and $b_0 \neq 0$ (using the notation after (2.2.2)).

We can now use the descriptions in Sects. 2.2.1 and 2.2.2 to write down explicit generators for each of R and U. Now $C_G(x) = U \rtimes R$ where $R \cong \mathrm{GL}_1(3) \times \mathrm{GL}_2(9) \times \mathrm{GL}_1(9) \times \mathrm{GL}_1(27)$ and U has order 3^γ where

$$\gamma = \sum_{i=1}^{h} d_i \left(2 \sum_{j<\ell} \lambda_{i,j} l_{i,j} l_{i,\ell} + \sum_{j} (\lambda_{i,j} - 1) l_{i,j}^2 \right) = 2[2(1\cdot2\cdot1)+(2-1)1^2] = 10.$$

Example 2.4.3 Consider $x \in \mathrm{SL}_8(3)$ where

$$x = \begin{pmatrix} 0&1&1&0&0&0&0&0\\ 1&2&0&1&0&0&0&0\\ 0&0&0&1&0&0&0&0\\ 0&0&1&2&0&0&0&0\\ 0&0&0&0&0&1&1&0\\ 0&0&0&0&1&2&0&1\\ 0&0&0&0&0&0&0&1\\ 0&0&0&0&0&0&1&2 \end{pmatrix}.$$

Its minimal polynomial is $(t^2 + t - 1)^2$. As usual, $C_{\mathrm{GL}_8(3)}(x) = U \rtimes R$ where $R \cong \mathrm{GL}_2(9)$. Every element of R can be identified with some $z_{1,1} \in \mathrm{GL}_2(9)$ and its determinant is $\mathrm{N}(\det z_{1,1})^2$, where N is the norm map $\mathbb{F}_9 \to \mathbb{F}_3$. For example, if ω is a root of $t^2 + t - 1$ in $\mathbb{F}_9$ and

$$z_{1,1} = \begin{pmatrix} \omega & 1 \\ \omega+2 & 0 \end{pmatrix},$$

then the corresponding element of R is

$$\begin{pmatrix} 0&1&0&0&1&0&0&0\\ 1&2&0&0&0&1&0&0\\ 0&0&0&1&0&0&1&0\\ 0&0&1&2&0&0&0&1\\ 2&1&0&0&0&0&0&0\\ 1&1&0&0&0&0&0&0\\ 0&0&2&1&0&0&0&0\\ 0&0&1&1&0&0&0&0 \end{pmatrix},$$

and has determinant $\mathrm{N}(\det z_{1,1})^2 = \mathrm{N}(1-\omega)^2 = 1$.

Example 2.4.4 Consider $x \in \mathrm{SL}_6(13)$ where

$$x = \begin{pmatrix} 1&1&0&0&0&0 \\ 0&1&1&0&0&0 \\ 0&0&1&0&0&0 \\ 0&0&0&1&1&0 \\ 0&0&0&0&1&1 \\ 0&0&0&0&0&1 \end{pmatrix}.$$

Its minimal polynomial is $(t-1)^3$. Here x has two Jordan blocks of order 3, and $d = \gcd(3, 3, 13-1) = 3$. Thus, the conjugacy class of x in $\mathrm{GL}_6(13)$ splits into three classes in $\mathrm{SL}_6(13)$. Observe 2 is a primitive element of $\mathbb{F}_{13}$. For $i = 0, 1, 2$,

$$y_i := \begin{pmatrix} 1 &&& \\ & \ddots && \\ && 1 & \\ &&& 2^i \end{pmatrix}.$$

Hence, the representatives for the three classes in $\mathrm{SL}_6(13)$ are

$$x^{y_0} = \begin{pmatrix} 1&1&0&0&0&0 \\ 0&1&1&0&0&0 \\ 0&0&1&0&0&0 \\ 0&0&0&1&1&0 \\ 0&0&0&0&1&1 \\ 0&0&0&0&0&1 \end{pmatrix}, \quad x^{y_1} = \begin{pmatrix} 1&1&0&0&0&0 \\ 0&1&1&0&0&0 \\ 0&0&1&0&0&0 \\ 0&0&0&1&1&0 \\ 0&0&0&0&1&2 \\ 0&0&0&0&0&1 \end{pmatrix}, \quad x^{y_2} = \begin{pmatrix} 1&1&0&0&0&0 \\ 0&1&1&0&0&0 \\ 0&0&1&0&0&0 \\ 0&0&0&1&1&0 \\ 0&0&0&0&1&4 \\ 0&0&0&0&0&1 \end{pmatrix}.$$

Chapter 3
Preliminaries on Classical Groups

In this chapter we first define the symplectic, orthogonal and unitary groups. Building on the work of Britnell [4] and Milnor [25], we then give in Theorem 3.2.6 necessary and sufficient criteria for $X \in \mathrm{GL}(V)$ to preserve a non-degenerate alternating, hermitian or quadratic form on V in terms of the elementary divisors of X.

3.1 The Finite Classical Groups

Let $F = \mathbb{F}_{q^u}$, where u is 1 or 2, and let $\bar{\lambda} = \lambda^q$ for $\lambda \in F$, so that $\lambda \mapsto \bar{\lambda}$ is a field automorphism of order u. Let $V = F^n$ be an n-dimensional vector space over F. For $X = (x_{ij}) \in \mathrm{GL}_n(F)$, define $\bar{X} = (\bar{x}_{ij})$, and set $X^* = \bar{X}^{\mathrm{t}}$, the transpose of $\bar{X}$. We also regard X as an element of $\mathrm{GL}(V)$, namely the map $v \mapsto Xv$; similarly for $\bar{X}$ and X^*.

We now define the classical symplectic, orthogonal and unitary groups as isometry groups of sesquilinear forms $\beta : V \times V \to F$ or quadratic forms $Q : V \to F$. We first recall a few notions about such forms. We consider sesquilinear forms $\beta : V \times V \to F$ such that

- β is left-linear: $\beta(au_1+bu_2, v) = a\beta(u_1, v)+b\beta(u_2, v)$ for $a, b \in F, u_i, v \in V$, and
- $\beta(v, u) = \overline{\beta(u, v)}$ for $u, v \in V$,

and quadratic forms $Q : V \to F$ such that

- $Q(av) = a^2 Q(v)$ for $a \in F$, $v \in V$, and
- the function $\beta_Q(u, v) = Q(u + v) - Q(u) - Q(v)$ is a bilinear form on V.

For a basis $B = \{v_1, \ldots, v_n\}$ of V, the matrix of a sesquilinear form β with respect to B is the $n \times n$ matrix $\beta_B = (\beta(v_i, v_j))$, and the matrix of a quadratic form

G. De Franceschi et al., *Conjugacy in Finite Classical Groups*, Springer Monographs in Mathematics, https://doi.org/10.1007/978-3-031-86461-2_3

Q is $A = (a_{ij})$, where $a_{ii} = Q(v_i)$, $a_{ij} = \beta_Q(v_i, v_j)$ for $i < j$, and $a_{ij} = 0$ for $i > j$. Note that $Q(v) = vAv^{\mathrm{t}}$ (writing vectors relative to the basis B), and $A + A^{\mathrm{t}}$ is the matrix of β_Q. We say that $g \in \mathrm{GL}(V)$ is an *isometry* of β if $\beta(ug, vg) = \beta(u, v)$ for all $u, v \in V$. Writing matrices with respect to a basis B, if $A = \beta_B$ is the matrix of β, then $X \in \mathrm{GL}_n(F)$ is an isometry if and only if $XAX^* = A$. Similarly, $g \in \mathrm{GL}(V)$ is an isometry of a quadratic from Q if $Q(vg) = Q(v)$ for all $v \in V$.

Two sesquilinear forms β_1, β_2 on V are *congruent* if there exists $T \in \mathrm{GL}(V)$ such that $\beta_1(uT, vT) = \beta_2(u, v)$ for all $u, v \in V$. If A_1 and A_2 are the matrices of β_1 and β_2 respectively, then this condition is equivalent to the existence of $S \in \mathrm{GL}_n(F)$ such that $SA_1\bar{S}^{\mathrm{t}} = A_2$. Similarly, quadratic forms Q_1, Q_2 are *congruent* if there exists $T \in \mathrm{GL}(V)$ such that $Q_1(vT) = Q_2(v)$ for all $v \in V$.

A sesquilinear form β is *non-degenerate* if its matrix β_B with respect to every basis B is invertible; and a quadratic form Q is non-degenerate if the associated bilinear form β_Q is non-degenerate. A subspace U of V is *totally singular* with respect to β (resp. Q) if β (resp. Q) vanishes on restriction to U. The *orthogonal complement* $U^\perp$ of U with respect to β (resp. Q) is $\{v \in V : \beta(u, v) = 0\ \forall u \in U\}$ (resp. $\{v \in V : \beta_Q(u, v) = 0\ \forall u \in U\}$).

We now define the classical groups; all assertions can be found in [30], for example.

- *Symplectic group* $\mathrm{Sp}(V)$: here $u = 1$, n is even and $\mathrm{Sp}(V)$ is the isometry group of a non-degenerate alternating bilinear form β on V. Up to congruence there is a unique such form. The corresponding matrix group is denoted $\mathrm{Sp}_n(q)$.
- *Unitary group* $\mathrm{GU}(V)$: here $u = 2$, and $\mathrm{GU}(V)$ is the isometry group of a non-degenerate hermitian form β on V. Up to congruence there is a unique such form. The corresponding matrix group is denoted $\mathrm{GU}_n(q)$.
- *Orthogonal group* $\mathrm{O}(V)$: here $u = 1$, and $\mathrm{O}(V)$ is the isometry group of a non-degenerate quadratic form Q on V. The associated symmetric bilinear form is β_Q, as defined above. If q is odd, then $\mathrm{O}(V)$ is also the isometry group of β_Q. Up to congruence there are two non-degenerate quadratic forms on V. If n is odd, then q is odd and the forms are Q and λQ, where λ is a non-square in F. When $n = 2k$, the congruence classes are distinguished by the dimension, k or $k - 1$, of a maximal totally singular subspace: if the dimension is k, then the orthogonal group is denoted $\mathrm{O}^+(V)$ or $\mathrm{O}^+_{2k}(q)$; if the dimension is $k - 1$, then it is denoted $\mathrm{O}^-(V)$ or $\mathrm{O}^-_{2k}(q)$. We denote the sign $\pm$ by $\mathrm{sgn}(Q)$, and say that Q is of *plus* or *minus type* accordingly. For $n = 2k$ and q odd, $\mathrm{sgn}(Q)$ is also determined by the *discriminant* $D(Q)$, which is the determinant of the matrix of β_Q modulo the squares $(F^*)^2$: namely, if $D(Q)$ is a square (resp., a non-square), then $\mathrm{sgn}(Q) = (-1)^{k(q-1)/2}$ (resp., $(-1)^{k(q+1)/2}$)—see [21, Prop. 2.5.10].

We often denote the isometry group of the sesquilinear form β by the symbol $C(\beta)$, and the isometry group of the quadratic form Q by $C(Q)$, sometimes replacing the form by its associated matrix.

The *special* unitary and orthogonal groups are the intersections of the isometry groups with $\mathrm{SL}(V)$, and denoted by $\mathrm{SU}(V) = \mathrm{GU}(V) \cap \mathrm{SL}(V)$ and $\mathrm{SO}(V) =$

$O(V) \cap SL(V)$. Finally, $SO(V)$ has a subgroup $\Omega(V)$ of index 2 which is the kernel of the *spinor norm* map $SO(V) \to C_2$ (see [21, §2.5]).

Generating sets of size 2 are recorded in [31] for the linear, symplectic, and unitary groups; generating sets of size at most 4 are recorded in [28] for the orthogonal groups. These *standard generators*, written with respect to a fixed form, are used to define (the standard copies of) the groups in MAGMA.

3.2 Membership of Classical Groups

Let $\mathcal{C} = \mathcal{C}(\beta)$ or $\mathcal{C}(Q)$ be a classical group on V, as defined in the previous section. The first step in classifying the conjugacy classes of $\mathcal{C}$ is to determine which classes of $GL(V)$ have elements in $\mathcal{C}$, and in this section we do this. The main result is Theorem 3.2.6. Our approach is modelled on those of Britnell [4, §5.1] and Milnor [25, §3].

Recall that $F = \mathbb{F}_{q^u}$, where u is 1 or 2, and $\lambda \mapsto \bar{\lambda}$ is a field automorphism of order u (where $\bar{\lambda} = \lambda^q$). For a monic polynomial $f(t) = t^d + a_{d-1}t^{d-1} + \cdots + a_0 \in F[t]$ with $a_0 \neq 0$, define $\bar{f}(t) = t^d + \bar{a}_{d-1}t^{d-1} + \cdots + \bar{a}_0$, and define the *dual polynomial* f^* of f by

$$f^*(t) = \bar{a}_0^{-1} t^d \bar{f}(t^{-1}). \tag{3.2.1}$$

Note that $(f^*)^* = f$ and $(fg)^* = f^*g^*$ for monic polynomials f and g. In particular, f is irreducible if and only if f^* is.

The next two results are essentially [4, Lemma 5.1 and Thm. 5.2]. They give information about the elementary divisors and the $F[t]$-module structure of V induced by an element of a classical group $\mathcal{C}(\beta)$ or $\mathcal{C}(Q)$.

Proposition 3.2.1 *Let β be a non-degenerate alternating, symmetric, or hermitian form on V, and let Q be a non-degenerate quadratic form. Let $X \in \mathcal{C}(\beta)$ or $\mathcal{C}(Q)$, and let U be an X-invariant subspace of V with $U \cap U^\perp = \{0\}$. If X has minimal polynomial f on U, then $f = f^*$.*

Proof First assume $X \in \mathcal{C}(\beta)$. Observe that $\beta(uX, v) = \beta(u, vX^{-1})$ for all $u, v \in U$. Write $f(t) = \sum_{i=0}^{d} a_i t^i$, where $a_0 \neq 0$ and $a_d = 1$. Now

$$\begin{aligned} 0 &= \beta(Uf(X), U) \\ &= \beta(U(\sum_i a_i X^i), U) \\ &= \beta(U, U\sum_i \overline{a_i} X^{-i}) \end{aligned}$$

$$= \beta(U, \overline{a_0}^{-1} U X^d \sum_i \overline{a_i} X^{-i})$$

$$= \beta(U, Uf^*(X)).$$

Since U is non-degenerate by hypothesis, the identity $\beta(U, Uf^*(X)) = 0$ implies $Uf^*(X) = 0$, so f divides f^*. But both f and f^* are monic of degree d, so $f = f^*$.

For $X \in \mathcal{C}(Q)$, the above proof goes through verbatim, with β_Q replacing β. □

Proposition 3.2.2 *Let β, Q be as in Proposition 3.2.1, and let $X \in \mathcal{C}(\beta)$ or $\mathcal{C}(Q)$. There exists an orthogonal decomposition $V = \bigoplus_i U_i$ such that, for each i, one of the following holds:*

(i) *X acts cyclically on U_i with minimal polynomial f^e for some $e \geq 1$, where f is irreducible and $f = f^*$;*
(ii) *$U_i = W \oplus W^*$ and X acts cyclically on W (resp. W^*) with minimal polynomial f^e (resp. f^{*e}) for some $e \geq 1$, where f is irreducible.*

Proof For notational convenience, in the case where $X \in \mathcal{C}(Q)$ we write β for the associated bilinear form β_Q.

Let U be one of the summands in the decomposition of V into cyclic X-submodules given in (2.1.1). Then X acts cyclically on U with minimal polynomial f^e, where f is irreducible and $e \geq 1$. Let W be an X-invariant subspace of V such that $V = U \oplus W$, and let $U^* = W^\perp$. Since β is non-degenerate,

$$U^* \cap U^\perp = U^\perp \cap W^\perp = (U \oplus W)^\perp = V^\perp = 0, \tag{3.2.2}$$

so for each non-zero $v \in U^*$ there exists $u \in U$ such that $\beta(u, v) \neq 0$. Moreover, for every non-zero $u \in U$, there exists $v \in U^*$ such that $\beta(u, v) \neq 0$ (otherwise $u \in U^{*\perp} = W$, a contradiction). We distinguish three cases.

Case 1: Assume $U \cap U^\perp = 0$. Then U is non-degenerate, so $f = f^*$ by Proposition 3.2.1; thus U is as in conclusion (i) of the proposition.

Case 2: Assume $U \cap U^\perp \neq 0$ and $U \cap U^* \neq 0$. Since $U \cap U^*$ is X-invariant, $Uf(X)^{e-1} \subseteq U \cap U^*$ (because X acts cyclically on U with minimal polynomial f^e). For the same reason $Uf(X)^{e-1} \subseteq U \cap U^\perp$. Hence $Uf(X)^{e-1} \subseteq U^* \cap U^\perp$, contradicting (3.2.2).

Case 3: Assume $U \cap U^\perp \neq 0$ and $U \cap U^* = 0$. We show that β is non-degenerate on $U \oplus U^*$: namely, for every $u \in U$, $v \in U^*$ with $u + v \neq 0$, there exists $z \in U \oplus U^*$ with $\beta(u + v, z) \neq 0$. If $v = 0$, then take $z \in U^*$ such that $\beta(u, z) \neq 0$; if $v \neq 0$ and $u \in U^\perp$, then take $z \in U$ such that $\beta(z, v) \neq 0$ (note that z exists in both cases since $U^\perp \cap U^* = 0$). Now suppose $v \neq 0$ and $u \notin U^\perp$. Let $a = \min\{n \in \mathbb{N} \mid uf(X)^n \in U^\perp\}$. By our choice of u, observe that $1 \leq a < e$. Consider

$$(u + v)f(X)^a = uf(X)^a + vf(X)^a.$$

If $uf(X)^a = 0$, then $uf(X)^{a-1} \in \ker f(X) = Uf(X)^{e-1} \subseteq U \cap U^\perp$, contradicting the minimality of a. So $uf(X)^a \neq 0$. If $vf(X)^a = 0$, then there exists $w \in U^*$ such that $\beta(uf(X)^a, w) \neq 0$, so $\beta((u+v)f(X)^a, w) \neq 0$, and we can choose $z = w\overline{f}(X^{-1})^a$. So we can assume that $vf(X)^a \neq 0$. Hence there exists $w \in U$ such that $\beta(w, v\overline{f}(X^{-1})^a) \neq 0$. Since $uf(X)^a \in U^\perp$, it follows that $\beta((u+v)f(X)^a, w) \neq 0$, so we can again choose $z = w\overline{f}(X^{-1})^a$. Thus β is non-degenerate on $U \oplus U^*$.

Let $U_1 = U \oplus U^*$. For every n,

$$\beta(Uf(X)^n, U^*) = 0 \;\Leftrightarrow\; \beta(U, U^* f^*(X)^n) = 0.$$

Since $U^* \cap U^\perp = 0$,

$$Uf(X)^n = 0 \;\Leftrightarrow\; U^* f^*(X)^n = 0.$$

This proves that the minimal polynomial of X on U^* is f^{*e}, and the action of X on U^* is cyclic because $\dim U = \dim U^*$. Hence conclusion (ii) holds for U. Since $V = U \oplus U^\perp$ and $U^\perp$ is X-invariant, we can repeat the argument for $U^\perp$, and the proposition follows by induction. □

Proposition 3.2.2 has the following consequence: if f^e is an elementary divisor of $X \in \mathcal{C}(\beta)$ or $\mathcal{C}(Q)$ of multiplicity m, then f^{*e} is also an elementary divisor of multiplicity m. Recall that $X, Y \in \mathrm{GL}(V)$ are *similar* if they are conjugate, in which case we write $X \sim Y$.

Corollary 3.2.3 *If X is an element of $\mathcal{C}(\beta)$ or $\mathcal{C}(Q)$, then X is similar to $\bar{X}^{-1}$.*

Proof By the definition of the dual polynomial, f^e is an elementary divisor of X if and only if f^{*e} is an elementary divisor of $\bar{X}^{-1}$. Hence, by the preceding remarks, X and $\bar{X}^{-1}$ have the same elementary divisors, and so they are similar. □

We now introduce the following notation, following [12].

Definition 3.2.4 Let $F = \mathbb{F}_{q^u}$ with $u = 1$ or 2, let $\lambda \mapsto \bar{\lambda}$ be a field automorphism of order u, and let f^* be as in (3.2.1). Define

$$\Phi_1 := \{f : f \in F[t] \mid f = f^* \text{ monic irreducible, } \deg f = 1\};$$

$$\Phi_2 := \{f : f \in F[t] \mid f = gg^*,\ g \neq g^*,\ g \text{ monic irreducible}\};$$

$$\Phi_3 := \{f : f \in F[t] \mid f = f^* \text{ monic irreducible, } \deg f > 1\}.$$

Let $\Phi := \Phi_1 \cup \Phi_2 \cup \Phi_3$ and let $f \in \Phi$. For $X \in \mathcal{C}(\beta)$ or $\mathcal{C}(Q)$ as above, and m a positive integer, f^m is a *generalized elementary divisor* of X if one of the following holds:

- $f \in \Phi_1 \cup \Phi_3$ and f^m is an elementary divisor of X;
- $f \in \Phi_2$, $f = gg^*$ and g^m is an elementary divisor of X (and so g^{*m} is as well).

Proposition 3.2.5 *The following hold:*

(i) *If* $u = 1$, *then* $\Phi_1 = \{t+1, t-1\}$.
(ii) *If* $u = 2$, *then* $\Phi_1 = \{t - \lambda : \lambda\bar{\lambda} = 1\}$.
(iii) *If* $u = 1$ *and* $f \in \Phi_3$, *then* $\deg f$ *is even.*
(iv) *If* $u = 2$ *and* $f \in \Phi_3$, *then* $\deg f$ *is odd.*

Proof Parts (i) and (ii) are clear.

Consider (iii). If $u = 1$ and $f \in \Phi_3$, then, for each root λ of f (in the splitting field), λ^{-1} is also a root, and $\lambda \neq \lambda^{-1}$ as $f \neq t \pm 1$. Hence $\deg f$ is even.

Finally consider (iv). Let $u = 2$ and $f \in \Phi_3$, and let $R(f)$ be the set of roots of f. The map $\alpha \mapsto \alpha^{-q}$ acts as a permutation on $R(f)$. Consider an orbit $\mathcal{O} = \{\lambda, \lambda^{-q}, \lambda^{q^2}, \ldots, \lambda^{(-q)^r}\}$ under this action, and subset $\mathcal{O}' = \{\lambda, \lambda^{q^2}, \lambda^{q^4}, \ldots\}$. The polynomial

$$g(t) = \prod_{\mu \in \mathcal{O}'} (t - \mu) = (t - \lambda)(t - \lambda^{q^2})(t - \lambda^{q^4}) \cdots$$

is a divisor of f and belongs to $\mathbb{F}_{q^2}[t]$ because its coefficients are fixed by the field automorphism $\alpha \mapsto \alpha^{q^2}$. Since f is irreducible, the only possibility is that $g = f$ and $\mathcal{O}' = R(f)$. It follows that $\mathcal{O}$ is the unique orbit in $R(f)$, and also that $\mathcal{O} = \mathcal{O}'$, which implies that $|\mathcal{O}|$ is odd. □

Now we state the main result of this section.

Theorem 3.2.6 *Let* $F = \mathbb{F}_{q^u}$ *with* $u = 1$ *or* 2, *and let* $\lambda \mapsto \bar{\lambda}$ *be an automorphism of* F *of order* u. *Let* V *be a vector space over* F *and let* $X \in \mathrm{GL}(V)$.

(i) *Suppose* $u = 2$. *There exists a non-degenerate hermitian form* β *on* V *such that* $X \in \mathcal{C}(\beta)$ *if and only if* $X \sim \bar{X}^{-1}$.
(ii) *Suppose* $u = 1$. *There exists a non-degenerate alternating form* β *on* V *such that* $X \in \mathcal{C}(\beta)$ *if and only if* $X \sim X^{-1}$ *and every elementary divisor* $(t \pm 1)^{2k+1}$ *of* X *with* $k \in \mathbb{N}$ *has even multiplicity.*
(iii) *Suppose* q *is odd and* $u = 1$.
   (a) *There exists a non-degenerate quadratic form* Q *on* V *such that* $X \in \mathcal{C}(Q)$ *if and only if* $X \sim X^{-1}$ *and every elementary divisor* $(t \pm 1)^{2k}$ *with* $k \in \mathbb{N}^+$ *has even multiplicity.*
   (b) *Suppose* (a) *holds for* X *and* $\dim V$ *is even. Then* $X \in \mathcal{C}(Q)$ *for* Q *of both plus and minus types if and only if* X *has at least one elementary divisor* $(t \pm 1)^{2k+1}$ *for some* $k \in \mathbb{N}$. *If this is not the case, then* $X \in \mathcal{C}(Q)$ *for* Q *of plus type (resp. minus type) if and only if* $\sum_{f,e} e \cdot m(f^e)$ *is even (resp. odd), where the sum runs over all* $f \in \Phi_3$, *and* $m(f^e)$ *is the multiplicity of* f^e *as an elementary divisor of* X.

(iv) *Suppose q is even and $u = 1$.*

(a) *There exists a non-degenerate quadratic form Q on V such that $X \in \mathcal{C}(Q)$ if and only if $X \sim X^{-1}$ and every elementary divisor $(t+1)^{2k+1}$ with $k \in \mathbb{N}$ has even multiplicity.*

(b) *Suppose* (a) *holds for X. Then $X \in \mathcal{C}(Q)$ for Q of both plus and minus types if and only if X has at least one elementary divisor $(t+1)^k$ for some $k \in \mathbb{N}^+$. If this is not the case, then $X \in \mathcal{C}(Q)$ for Q of plus type (resp. minus type) if and only if $\sum_{f,e} e \cdot m(f^e)$ is even (resp. odd), where the sum runs over all $f \in \Phi_3$, and $m(f^e)$ is the multiplicity of f^e as an elementary divisor of X.*

To prove the theorem we require several preliminary lemmas. The first is [20, 2.2]; we include a proof for completeness.

Lemma 3.2.7 *Suppose q is odd and X acts cyclically on V with minimal polynomial $(t-\epsilon)^{2k}$, where $\epsilon = 1$ or -1. Then there is no non-degenerate symmetric bilinear form β on V such that $X \in \mathcal{C}(\beta)$.*

Proof We prove this for $\epsilon = 1$; the proof for $\epsilon = -1$ is similar. Put $m = 2k$ and choose a basis $v_1, \ldots, v_m$ for V such that $v_1 X = v_1$ and $v_i X = v_{i-1} + v_i$ for all $i = 2, \ldots, m$. Suppose for a contradiction that β is a non-degenerate symmetric form on V such that $X \in \mathcal{C}(\beta)$. Since $\langle v_1 \rangle^\perp$ is an invariant X-submodule of V of dimension $m-1$, it must equal the unique submodule of this dimension, so $\langle v_1 \rangle^\perp = \langle v_1, \ldots, v_{m-1} \rangle$. Hence $\beta(v_1, v_m) \neq 0$. Now,

$$\begin{aligned} 0 \neq \beta(v_1, v_m) &= \beta(v_1, v_1(X-1)^{m-1}) \\ &= \beta(v_1(X^{-1}-1)^{m-1}, v_1) \\ &= \beta((-1)^{m-1} v_1 X^{-m+1}(X-1)^{m-1}, v_1) \\ &= (-1)^{m-1} \beta(v_m X^{-m+1}, v_1) \\ &= (-1)^{m-1} \beta(v_m, v_1) \\ &= -\beta(v_1, v_m). \end{aligned}$$

This is impossible in odd characteristic. □

The next lemma also follows from [20].

Lemma 3.2.8 *Let q be odd. Let β be a non-degenerate symmetric form on V, and let $X \in \mathcal{C}(\beta)$. Let U be a non-degenerate direct summand of $V \downarrow X$ that satisfies conclusion* (ii) *of Proposition 3.2.2. Then one of the following holds:*

(i) *there are non-degenerate X-submodules U_1, U_2 of U of dimension $\frac{1}{2} \dim U$ such that $U = U_1 + U_2$, an orthogonal sum;*

(ii) *there exist totally isotropic X-submodules W, W^* of U such that $U = W \oplus W^*$, and either $f \neq f^*$ or $f^e = (t \pm 1)^{2k}$.*

Proof If $f \neq f^*$, then it follows directly from Lemma 6.1.4 (whose proof is independent of this chapter) that (ii) holds. Suppose that $f = f^*$ and (i) does not hold. Then, in the terminology of [20, 1.8], U is of "Type 1". It follows from [20, 2.1] that $f = t \pm 1$ and e is even; moreover, [20, 2.4] gives the existence of totally isotropic X-submodules W, W^* such that $U = W \oplus W^*$. Thus (ii) holds. □

Lemma 3.2.9 *Suppose that $u = 1$, so $F = \mathbb{F}_q$. Let $f \in F[t]$ be an irreducible polynomial with $f = f^*$ and $d = \deg f > 1$, and let X be its companion matrix over F. If Q is a non-degenerate quadratic form on $V = F^d$ such that $X \in \mathcal{C}(Q)$, then Q has minus type.*

Proof Note that d is even by Proposition 3.2.5. We show in Theorem 6.1.2 (proved independently of this chapter) that $C_{\mathcal{C}(Q)}(X) \cong \mathrm{GU}_1(q^{d/2})$, and so it is a subgroup of $\mathcal{C}(Q) \cong \mathrm{O}^{\epsilon}_d(q)$, where $\epsilon = \pm$. But $|\mathrm{GU}_1(q^{d/2})| = 1 + q^{d/2}$ divides $|\mathrm{O}^+_d(q)|$ only if $(d, q) = (2, 3)$ or $(6, 2)$. In these cases the only possible polynomials f are $t^2 + 1 \in \mathbb{F}_3[t]$ and $t^6 + t^3 + 1 \in \mathbb{F}_2[t]$; we check that their companion matrices do not preserve non-degenerate quadratic forms of plus type. Hence $\epsilon = -$. □

Proof of Theorem 3.2.6

(1) We begin by proving the left to right implications (the "only if" assertions) in parts (i), (ii), (iii)(a) and (iv)(a) of the theorem. Suppose that $X \in \mathcal{C}(\beta)$ or $\mathcal{C}(Q)$. Then $X \sim \bar{X}^{-1}$ by Corollary 3.2.3. Moreover, if the multiplicity of $(t \pm 1)^{2k+1}$ as an elementary divisor of X is odd for some k, then, by Proposition 3.2.2, there must be a non-degenerate subspace U_1 on which X acts cyclically with minimal polynomial $(t \pm 1)^{2k+1}$; in particular, $\dim U_1 = 2k + 1$ is odd, and so β cannot be alternating. This completes the proof of the left to right implication in (ii) and (iv)(a). Similarly, if q is odd and the multiplicity of $(t \pm 1)^{2k}$ is odd for some k, then β cannot be symmetric, by Lemma 3.2.7, completing the left to right part of (iii)(a).

(2) Next we prove the right to left implications (the "if" assertions) in parts (i), (ii), (iii)(a) and (iv)(a) of the theorem. Suppose that $X \sim \bar{X}^{-1}$ and X satisfies the condition on elementary divisors $(t \pm 1)^{2k+1}$ in parts (ii) and (iv)(a), and on elementary divisors $(t \pm 1)^{2k}$ in part (iii)(a). We can assume that X has a unique generalized elementary divisor f^e with $f \in \Phi$, and that this has multiplicity $m = 1$ or 2, where $m = 2$ only in the cases where even multiplicity is assumed in (ii), (iii)(a) and (iv)(a); once we exhibit forms β or Q fixed by such elements, we can just take direct sums to exhibit forms fixed by X in the general case.

Case a: $f \in \Phi_1$. Here X has a unique elementary divisor $(t - \lambda)^e$, where $\lambda\bar{\lambda} = 1$. Hence $X = \lambda u$, where u is unipotent. Under the assumed conditions on the multiplicity m, unipotent elements are defined in $\mathcal{C}(\beta)$ and $\mathcal{C}(Q)$ in the following sections: Sect. 4.1.3 for unitary groups; Sects. 4.1.1 and 4.1.2 for symplectic and orthogonal groups in odd characteristic; and Sect. 5.1 for symplectic and orthogonal groups in characteristic 2.

Case b: $f \in \Phi_2$. Here $f = gg^*$, where g is irreducible and $g \neq g^*$, and f^e has multiplicity 1 as a generalized elementary divisor of X. Let $d = \deg g$. In an appropriate basis, X has block diagonal matrix $X = Y \oplus Y^{*-1}$, with Y a Jordan block relative to g^e. Then X is an isometry for the form with matrix B or, in the orthogonal case, for the quadratic form with matrix A, where

$$B = \begin{pmatrix} \mathbb{O} & \mathbb{I} \\ \varepsilon\mathbb{I} & \mathbb{O} \end{pmatrix}, \quad A = \begin{pmatrix} \mathbb{O} & \mathbb{I} \\ \mathbb{O} & \mathbb{O} \end{pmatrix},$$

$\mathbb{O}$ is the $de \times de$ zero matrix, $\mathbb{I}$ is the $de \times de$ identity matrix, and $\varepsilon = -1$ in the symplectic case and 1 otherwise.

This completes the argument for Case **b**. Note that $V = U \oplus W$, where X acts cyclically on U (resp. W) with minimal polynomial g^e (resp. g^{*e}), and U and W are totally singular.

Case c: $f \in \Phi_3$. We extend [4, Thm. 5.4] to unitary groups and to symplectic and orthogonal groups in even characteristic. In this case X has a unique elementary divisor f^e with multiplicity 1, where $f = f^*$ is irreducible of degree $d > 1$. Then X acts cyclically on V with minimal polynomial f^e. Let U_1 and U_2 be isomorphic to V (as X-modules) and write $U = U_1 \oplus U_2$. Using the same argument as in Case **b**, we see that there exists a non-degenerate form β (or quadratic form Q) on U such that $X \in \mathcal{C}(\beta)$ and U_1, U_2 are totally singular.

Now suppose there exists $v \in U$ and $i \in \mathbb{Z}$ such that $\beta(v, vX^i f(X)^{e-1}) \neq 0$. Let W be the cyclic X-submodule of U generated by v. The minimal polynomial of X on W is f^c for some $c \leq e$. If $c < e$, then $vf(X)^{e-1} = 0$, so $\beta(v, vX^i f(X)^{e-1}) = \beta(v, 0) = 0$, a contradiction. Hence $c = e$, and the only submodules of W are of the form $Wf(X)^m$ for some $m \leq e$. Moreover, W is non-degenerate: if not, then $W \cap W^\perp$ is a nonzero submodule of W, hence it equals $Wf(X)^m$ for some $m < e$; but this implies that $vf(X)^{e-1} \in W^\perp$, a contradiction. Thus W is non-degenerate and X is an element of $\mathcal{C}(\beta_W)$ or $\mathcal{C}(Q_W)$, where β_W, Q_W are the restrictions of β, Q to W. Since W is isomorphic to V as X-module, such a form must exist on V, as required.

By the previous paragraph, the proof for the case $f \in \Phi_3$ is complete except when $\beta(v, vX^i f^{e-1}(X)) = 0$ for every $v \in V$ and $i \in \mathbb{Z}$. Suppose for a contradiction that this is the case. For $v = v_1 + v_2$ with $v_i \in U_i$, this condition is equivalent to

$$\beta(v_1, v_2X^i f(X)^{e-1}) + \beta(v_2, v_1X^i f(X)^{e-1}) = 0 \tag{3.2.3}$$

for all $v_1 \in U_1$, $v_2 \in U_2$ and $i \in \mathbb{Z}$. In the unitary case, choosing $\lambda \in \mathbb{F}_{q^2}$ such that $\lambda \neq \bar{\lambda}$, and replacing v_1 by λv_1 in (3.2.3) gives

$$\lambda\beta(v_1, v_2X^i f(X)^{e-1}) + \bar{\lambda}\beta(v_2, v_1X^i f(X)^{e-1}) = 0.$$

Combined with (3.2.3), and taking $i = 0$, this implies that $\beta(v_1, v_2 f(X)^{e-1}) = 0$ for all $v_1 \in U_1$, $v_2 \in U_2$, which implies that the restriction $f(X)^{e-1} \downarrow U_2 = 0$, a contradiction.

Now consider the symplectic and orthogonal cases. By a straightforward computation, (3.2.3) implies the following sequence of identities:

$$\beta(v_1, v_2[X^i f(X)^{e-1} + \varepsilon X^{-i} f(X^{-1})^{e-1}]) = 0,$$
$$\beta(v_1, v_2[X^i f(X)^{e-1} + \varepsilon f(0)^{e-1} X^{-i-d(e-1)} f(X)^{e-1}]) = 0, \quad \text{(using } f = f^*)$$
$$\beta(v_1, v_2 f(X)^{e-1}[X^i + \varepsilon f(0)^{e-1} X^{-i-d(e-1)}]) = 0,$$
$$\beta(v_1, v_2 f(X)^{e-1}[X^{d(e-1)+2i} + \varepsilon f(0)^{e-1}]) = 0.$$

Since β is non-degenerate on $U_1 \oplus U_2$, for every $i \in \mathbb{Z}$, the restriction

$$f(X)^{e-1}\left(X^{d(e-1)+2i} + \varepsilon f(0)^{e-1}\right) \downarrow U_2 = 0.$$

This implies that $f(t)$ divides $t^{d(e-1)+2i} + \varepsilon f(0)^{e-1}$ for every i such that $d(e-1) + 2i$ is non-negative. Since we are in the symplectic or orthogonal case, d is even by Proposition 3.2.5. Hence, choosing $i = 1 - d(e-1)/2$, we see that $f(t)$ divides $t + \varepsilon f(0)^{e-1}$ of degree 1, which is impossible.

This completes the argument for Case **c**. Parts (i), (ii), (iii)(a) and (iv)(a) of the theorem are now established.

(3) It remains to prove parts (iii)(b) and (iv)(b) of the theorem. Consider (iii)(b). We suppose (as in (iii)(a)) that $\dim V$ is even, $X \sim X^{-1}$, and every elementary divisor $(t \pm 1)^{2k}$ with $k \in \mathbb{N}^+$ has even multiplicity. Combining Propositions 3.2.2 and 3.2.8, we see that $V \downarrow X$ has an orthogonal decomposition $V = \bigoplus_i U_i$ such that, for each i, one of the following holds:

(a) X acts cyclically on U_i with minimal polynomial f^e for some $e \geq 1$, where f is irreducible and $f = f^*$;

(b) $U_i = W \oplus W^*$ and X acts cyclically on W (resp. W^*) with minimal polynomial f^e (resp. f^{*e}) for some $e \geq 1$, where f is irreducible; moreover, W and W^* are totally isotropic, and either $f \neq f^*$ or $f^e = (t \pm 1)^{2k}$.

Assume first that X has an elementary divisor $(t \pm 1)^{2k+1}$ for some k. Then there must be a summand U_i on which X acts cyclically with minimal polynomial $(t \pm 1)^{2k+1}$. From Sect. 4.1.2, we see that there are quadratic forms on U_i of both square and non-square discriminants preserved by the action of X (corresponding to the unipotent elements $V_1(2k+1)$ and $V_\alpha(2k+1)$). Hence X preserves quadratic forms of both square and non-square discriminants on V, showing that $X \in C(Q)$ for Q of both plus and minus types, as required.

Now assume that X has no elementary divisor $(t \pm 1)^{2k+1}$. Consider a summand U_i. If U_i is as in **(b)** above, then clearly $X^{U_i} \in \mathrm{O}^+(U_i)$. Now suppose U_i is as in **(a)**. Here $f = f^*$ by Proposition 3.2.1. If $f = t \pm 1$, then $f^e = (t \pm 1)^{2k}$ for some k, by our assumption at the beginning of this paragraph. But this contradicts Lemma 3.2.7. Hence $\deg f \geq 2$ and $f \in \Phi_3$. Now X acts cyclically on U_i with minimal polynomial f^e. Let $X = SU$ be the Jordan decomposition of X, where S is semisimple and U unipotent. An application of Lemma 3.2.8 now shows that V is an orthogonal sum of cyclic S-submodules $U'_1, \ldots, U'_e$, on which S acts cyclically with minimal polynomial f. By Lemma 3.2.9, each U'_i has minus type, and hence U_i has type $(-1)^e$. It follows that V has type $(-1)^l$, where $l = \sum_{f,e} e \cdot m(f^e)$ and the sum runs over $f \in \Phi_3$.

This concludes the proof of part (iii)(b) of the theorem. The proof of (iv)(b) is very similar, and in fact easier, as we can work as above with the decomposition $V = \bigoplus U_i$ of Proposition 3.2.2.

The proof of Theorem 3.2.6 is now complete. □

Chapter 4
Unipotent Classes in Good Characteristic

As described in Chap. 1, our approach to the conjugacy problems (1)–(3) of Sect. 1.1 splits naturally into the analysis of the unipotent and semisimple classes. In this chapter and the next we solve the conjugacy problems for unipotent classes in classical groups. Since we deal separately with the different families of classical groups, we drop the $\mathcal{C}(\beta)$ and $\mathcal{C}(Q)$ notation of the previous chapter, and let G be $\mathrm{Sp}_{2n}(q)$, $\mathrm{O}_n^\epsilon(q)$, or $\mathrm{GU}_n(q)$ where $q = p^a$ with p prime. Recall that the characteristic p is *bad* if G is symplectic or orthogonal and $p = 2$; otherwise p is *good*. We address the classical groups in good characteristic in this chapter and the bad characteristic cases in Chap. 5.

4.1 Unipotent Class Representatives

The unipotent class representatives of the finite classical groups are given (in all positive characteristics) in [14]. For effective computation with these classes in good characteristic, we choose representatives that differ from those of [14]. In this section we present these representatives and structural information about their centralizers.

As outlined in Sect. 1.2, the choice of representatives and the construction of their centralizers is based on theory developed in [22]. We now present a more detailed summary of this theory. Let K be an algebraically closed field of odd characteristic p, and let $\bar{V}$ be an n-dimensional vector space over K. Let $(\ ,\ )$ be a non-degenerate symplectic or orthogonal bilinear form on $\bar{V}$, with isometry group $\bar{G} = \mathrm{Sp}(\bar{V})$ or $\mathrm{O}(\bar{V})$. The Lie algebra of $\bar{G}$ is

$$L(\bar{G}) = \{A \in \mathrm{End}(\bar{V}) : (vA, w) + (v, wA) = 0\ \forall v, w \in \bar{V}\}$$

G. De Franceschi et al., *Conjugacy in Finite Classical Groups*, Springer Monographs in Mathematics, https://doi.org/10.1007/978-3-031-86461-2_4

(see [22, Lemma 2.7]). By [22, Lemma 2.15], if $e \in L(\bar{G})$ is a nilpotent element, then $u := (1-e)(1+e)^{-1}$ is a unipotent element of $\bar{G}$, and the map

$$e \mapsto (1-e)(1+e)^{-1}$$

is a bijection, known as the *Cayley map*, from the set of nilpotent elements of $L(\bar{G})$ to the set of unipotent elements of $\bar{G}$. The Cayley map χ is $\bar{G}$-*equivariant*: namely, $\chi(e^g) = \chi(e)^g$ for $g \in \bar{G}$. In particular, $C_{\bar{G}}(u) = C_{\bar{G}}(e)$.

In [22, §3.3.2], various nilpotent elements e of $L(\bar{G})$ are defined, and the corresponding unipotent elements $u = (1-e)(1+e)^{-1}$ are shown to form a complete set of unipotent class representatives for $\bar{G}$. For each such nilpotent element e, a 1-dimensional *torus* $T = \{T(c) : c \in K^*\} \leq \bar{G}$ is constructed, with the property that $eT(c) = c^2 e$ for all $c \in K^*$. The space $\bar{V}$ is a direct sum of T-weight spaces $\bar{V}_{(i)}$ (for $i \in \mathbb{Z}$), where

$$\bar{V}_{(i)} = \{v \in \bar{V} : vT(c) = c^i v\}.$$

If we set $\bar{U}_{(i)} = \langle \bar{V}_{(j)} : j \geq i \rangle$, then the subspaces $\bar{U}_{(i)}$ form a flag

$$\cdots \subseteq \bar{U}_{(i+1)} \subseteq \bar{U}_{(i)} \subseteq \bar{U}_{(i-1)} \subseteq \cdots .$$

The stabilizer in $\bar{G}$ of this flag is a parabolic subgroup P, and the Levi decomposition of this parabolic is $P = QL$, with unipotent radical Q and Levi factor $L = C_{\bar{G}}(T)$. By [22, Prop. 3.10 and Lemma 3.13], $C_{\bar{G}}(u) = C_{\bar{G}}(e) = UR$, where U is the unipotent radical and $R = C_L(e)$ is reductive. Moreover, by [22, Lemma 2.29], $U = C_Q(e)$, and hence

$$C_{\bar{G}}(u) = C_{\bar{G}}(e) = C_Q(e)\, C_L(e). \tag{4.1.1}$$

The structure of the reductive group $R = C_L(e)$ and the dimension of the unipotent group $C_Q(e)$ are given by [22, Prop. 3.7, (3.4), (3.5)].

This summarizes the theory of unipotent class representatives and centralizers in symplectic and orthogonal groups over algebraically closed fields of odd characteristic. Over finite fields, the theory is similar, but requires various tweaks, as we now discuss.

4.1.1 Symplectic Groups

Let $G = \mathrm{Sp}_{2n}(q)$ with $q = p^a$ and p an odd prime, and let $(\ ,\ )$ be a symplectic form on $V = \mathbb{F}_q^n$ preserved by G. Let J_i be a unipotent $i \times i$ Jordan block matrix. By Theorem 3.2.6, the Jordan form of a unipotent element of G has an even number of blocks of each odd size, so it is a block diagonal sum

$$\bigoplus_{i=1}^{r} J_{2k_i}^{[a_i]} \oplus \bigoplus_{i=1}^{s} J_{2l_i+1}^{[2b_i]}, \tag{4.1.2}$$

where $k_1, \dots, k_r$ are distinct, as are $l_1, \dots, l_s$. By [14, Prop. 2.3], the number of conjugacy classes in G of elements with such a Jordan form is 2^r.

We now present representatives for these classes. While our representatives differ from those in [14], they are labelled in the same way. The finite symplectic Lie algebra is

$$L(G) = sp_{2n}(q) = \{A \in \mathrm{End}(V) : (vA, w) + (v, wA) = 0 \ \forall v, w \in V\}.$$

As discussed above, our representatives are images of nilpotent elements in $L(G)$ under the Cayley map.

Even Blocks $V_\beta(2m)$

Let α be a fixed non-square in $\mathbb{F}_q^*$, and let $\beta \in \{1, \alpha\}$. Let V_{2m} be a $2m$-dimensional vector space over $\mathbb{F}_q$ with basis $v_{-(2m-1)}, v_{-(2m-3)}, \dots, v_{2m-1}$. Define a symplectic form on V_{2m} by setting $(v_i, v_{-i}) = (-1)^{\lfloor i/2 \rfloor}$ for all i, and all other values $(v_i, v_j) = 0$. Define $e_\beta \in \mathrm{End}(V_{2m})$ by

$$\begin{aligned} e_\beta : v_i &\mapsto v_{i+2} \ (i < 2m-1,\ i \neq -1), \\ v_{-1} &\mapsto \beta v_1, \\ v_{2m-1} &\mapsto 0. \end{aligned}$$

Now e_β is a nilpotent element of $sp_{2m}(q)$. With respect to the given basis, e_β has matrix

$$\begin{pmatrix} 0 & 1 & & & & & & & \\ & 0 & 1 & & & & & & \\ & & \ddots & & & & & & \\ & & & 0 & 1 & & & & \\ & & & & 0 & \beta & & & \\ & & & & & 0 & 1 & & \\ & & & & & & \ddots & & \\ & & & & & & & 0 & 1 \\ & & & & & & & & 0 \end{pmatrix}.$$

We use the Cayley map to define a unipotent element

$$V_\beta(2m) = (1 - e_\beta)(1 + e_\beta)^{-1} \in G = \mathrm{Sp}(V_{2m}).$$

Observe that $V_\beta(2m)$ is a single Jordan block. Its matrix relative to the given basis is

$$\begin{pmatrix}
1 & -2 & 2\cdots & 2\epsilon & -2\epsilon\beta & 2\epsilon\beta & & \cdots & -2\beta \\
 & 1 & -2\cdots & -2\epsilon & 2\epsilon\beta & -2\epsilon\beta & & \cdots & 2\beta \\
 & & 1 & & & & & & \\
 & & & \ddots & & & & & \\
 & & & 1 & -2\beta & 2\beta & & \cdots & 2\epsilon\beta \\
 & & & & 1 & -2 & 2 & \cdots & 2\epsilon \\
 & & & & & 1 & -2 & \cdots & -2\epsilon \\
 & & & & & & & \ddots & \\
 & & & & & & & & 1
\end{pmatrix}$$

where $\epsilon = (-1)^{m+1}$ and the first β appears in the $(m+1)^{th}$ column. Writing $\bar{\mathbb{F}}_q$ for the algebraic closure of $\mathbb{F}_q$, define a 1-dimensional torus $T_1 = \{T_1(c) : c \in \bar{\mathbb{F}}_q^*\} \leq \mathrm{Sp}(\bar{V}_{2m})$, where $\bar{V}_{2m} = V_{2m} \otimes \bar{\mathbb{F}}_q$, by the action

$$T_1(c) : v_i \mapsto c^i v_i \text{ for all } i.$$

Note that $e_\beta T_1(c) = c^2 e_\beta$. Also one checks that $V_\beta(2m)$ is conjugate to the element labelled $V_{2\beta}(2m)$ in [14, §2.4].

Odd Blocks $W(2m+1)$

Let V_{4m+2} be a $(4m+2)$-dimensional space over $\mathbb{F}_q$ with basis

$$w_{-2m}, x_{-2m}, w_{-(2m-2)}, x_{-(2m-2)}, \ldots, w_{2m}, x_{2m},$$

and symplectic form defined by $(w_i, x_{-i}) = (-1)^{\lfloor i/2 \rfloor}$ for all i and all other values (w_i, w_j), (w_i, x_j), (x_i, x_j) equal 0. Define $e \in \mathrm{End}(V_{4m+2})$ by

$$\begin{aligned}
e : w_i &\mapsto w_{i+2} \ (i < 2m), \\
x_i &\mapsto x_{i+2} \ (i < 2m), \\
w_{2m} &\mapsto 0, \\
x_{2m} &\mapsto 0.
\end{aligned}$$

Now e is a nilpotent element of $sp_{4m+2}(q)$. Define a unipotent element

$$W(2m+1) = (1-e)(1+e)^{-1} \in G = \mathrm{Sp}(V_{4m+2}).$$

Observe that $W(2m+1)$ has Jordan form $J_{2m+1}^{[2]}$. Its matrix relative to the given basis is

$$\begin{pmatrix} I & -2I & 2I & \cdots & (-1)^{m+1}2I \\ & I & -2I & \cdots & (-1)^m 2I \\ & & & \ddots & \\ & & & & I \end{pmatrix}$$

where I denotes the 2×2 identity matrix. Define a 1-dimensional torus $T_1 \leq \mathrm{Sp}(\bar{V}_{4m+2})$ by the action

$$T_1(c) : w_i \mapsto c^i w_i,\ x_i \mapsto c^i x_i \quad \text{for all } i.$$

Class Representatives

By [14, Prop. 4.3], for a given Jordan form (4.1.2), there are 2^r conjugacy class representatives in $\mathrm{Sp}(V)$, namely

$$\bigoplus_{i=1}^{r} (V_{\beta_i}(2k_i) \oplus V_1(2k_i)^{[a_i-1]}) \oplus \bigoplus_{i=1}^{s} W(2l_i+1)^{[b_i]}, \tag{4.1.3}$$

where $V = \bigoplus_{i=1}^r V_{2k_i}^{[a_i]} \oplus \bigoplus_{i=1}^s V_{4l_i+2}^{[b_i]}$, and each $\beta_i \in \{1, \alpha\}$ (and the k_i are distinct, as are the l_i). As defined in the above notation, each V_{2k_i} (resp. V_{4l_i+2}) is a subspace of V on which the unipotent element $V_{\beta_i}(2k_i)$ or $V_1(2k_i)$ (resp. $W(2l_i+1)$) acts. Define T to be the 1-dimensional torus acting as $T_1(c)$ on each subspace $\bar{V}_{2k_i}$, $\bar{V}_{4l_i+2}$.

To define corresponding matrix representatives, choose an ordered basis of V according to the T-weights. Namely, list the spaces V_{2k_j}, V_{4l_j+2} in decreasing order of Jordan block sizes. For the ith space, if it is $V_\beta(2k)$, then label its basis as

$$v_{-(2k-1)}^{(i)}, v_{-(2k-3)}^{(i)}, \ldots, v_{2k-1}^{(i)},$$

and if it is $W(2l+1)$, then label its basis as

$$w_{-2l}^{(i)}, x_{-2l}^{(i)}, w_{-(2l-2)}^{(i)}, x_{-(2l-2)}^{(i)}, \ldots, w_{2l}^{(i)}, x_{2l}^{(i)}.$$

Now take a basis of V consisting of these vectors $v_j^{(i)}, w_j^{(i)}, x_j^{(i)}$ in increasing order lexicographically with respect to the pairs (j, i). Choose the matrix representative corresponding to the linear map (4.1.3) to be its matrix with respect to this ordered basis.

Example 4.1.1 Consider the unipotent element

$$V_\alpha(4) \oplus W(3) \oplus V_1(2) \in \mathrm{Sp}(V),$$

where $\dim V = 12$. The ordered basis is

$$v_{-3}^{(1)}, w_{-2}^{(2)}, x_{-2}^{(2)}, v_{-1}^{(1)}, v_{-1}^{(3)}, w_0^{(2)}, x_0^{(2)}, v_1^{(1)}, v_1^{(3)}, w_2^{(2)}, x_2^{(2)}, v_3^{(1)},$$

and the corresponding matrix representative is

$$\begin{pmatrix}
1 & & & -2 & & & & 2\alpha & & & & -2\alpha \\
 & 1 & & & & -2 & & & & 2 & & \\
 & & 1 & & & & -2 & & & & 2 & \\
 & & & 1 & & & & -2\alpha & & & & 2\alpha \\
 & & & & 1 & & & & -2 & & & \\
 & & & & & 1 & & & & -2 & & \\
 & & & & & & 1 & & & & -2 & \\
 & & & & & & & 1 & & & & -2 \\
 & & & & & & & & 1 & & & \\
 & & & & & & & & & 1 & & \\
 & & & & & & & & & & 1 & \\
 & & & & & & & & & & & 1
\end{pmatrix}.$$

Let $V_{(i)}$ denote the *T-weight space* for weight i: the span of all vectors in the basis with subscript i. Set $U_{(i)} = \langle V_{(j)} : j \geq i \rangle$—so $U_{(3)} = \langle v_3^{(1)} \rangle$, $U_{(2)} = \langle w_2^{(2)}, x_2^{(2)}, v_3^{(1)} \rangle$ and so on. The subspaces $U_{(i)}$ form a flag

$$U_{(3)} \subset U_{(2)} \subset U_{(1)} \subset \cdots \subset U_{(-3)} = V,$$

and the stabilizer in $G = \mathrm{Sp}(V)$ of this flag is a parabolic subgroup $P = QL$ with unipotent radical Q and Levi subgroup $L = C_G(T) = \mathrm{GL}_1(q) \times \mathrm{GL}_2(q) \times \mathrm{GL}_2(q) \times \mathrm{Sp}_2(q)$.

Returning to the general case, let P be the parabolic subgroup stabilizing the flag of V consisting of subspaces $U_{(i)} = \langle V_{(j)} : j \geq i \rangle$, where $V_{(i)}$ is the T-weight space for weight i. As in the example, $P = QL$, with unipotent radical Q and Levi subgroup $L = C_G(T)$. Now the discussion around (4.1.1) (the finite version also uses [22, Thm. 7.1]) gives the following result.

Theorem 4.1.2 *Let $G = \mathrm{Sp}_{2n}(q)$ with q odd, and let $u \in G$ be the unipotent matrix representative defined above for the linear map*

$$\bigoplus_{i=1}^{r}(V_{\beta_i}(2k_i) \oplus V_1(2k_i)^{[a_i-1]}) + \bigoplus_{i=1}^{s} W(2l_i+1)^{[b_i]}.$$

Let $P = QL$ be the corresponding parabolic subgroup.

(i) $C_G(u) = C_Q(u)C_L(u)$;
(ii) $C_L(u) \cong \prod_{i=1}^{s} \mathrm{Sp}_{2b_i}(q) \times \prod_{i=1}^{r} \mathrm{O}_{a_i}^{\epsilon_i}(q)$;
(iii) *rewrite the Jordan form of u as $\bigoplus J_i^{r_i}$; then $|C_Q(u)| = q^m$ where*

$$m = \frac{1}{2}\sum_i (i-1)r_i^2 + \sum_{i<j} i r_i r_j + \frac{1}{2}\sum_{i \text{ even}} r_i.$$

In part (ii), the sign $\epsilon_i = \pm$ is the sign of the orthogonal form with Gram matrix $\mathrm{diag}(\beta_i, 1, \ldots, 1)$. In Sect. 4.2.1 we justify this, and show how to construct the subgroups $C_Q(u)$ and $C_L(u)$ of the centralizer.

4.1.2 Orthogonal Groups

Let V be a vector space of dimension n over $\mathbb{F}_q$, where $q = p^a$ is odd, and let $(\, ,\,)$ be a non-degenerate symmetric bilinear form on V. Let $G = \mathrm{O}(V)$ be the corresponding orthogonal group. If n is even, then $G \cong \mathrm{O}_n^{\epsilon}(q)$ with $\epsilon = \pm$; if n is odd, then there is only one type of orthogonal group $\mathrm{O}_n(q)$, but for ease of notation we sometimes denote this by $\mathrm{O}_n^{\epsilon}(q)$. The Jordan form of a unipotent element of G has an even number of blocks of each even size, so it is

$$\bigoplus_{i=1}^{r} J_{2k_i+1}^{[a_i]} \oplus \bigoplus_{i=1}^{s} J_{2l_i}^{[2b_i]}, \tag{4.1.4}$$

where $k_1, \ldots, k_r$ are distinct, as are $l_1, \ldots, l_s$. By [14, Prop. 2.4], the number of conjugacy classes in G of elements with such a Jordan form is as follows:

- n odd: 2^{r-1} classes;
- n even: 2^{r-1} classes in each of $\mathrm{O}_n^+(q)$ and $\mathrm{O}_n^-(q)$, with the exception that if $r = 0$, then there is just one class in $\mathrm{O}_n^+(q)$ and none in $\mathrm{O}_n^-(q)$.

To produce class representatives, we use the finite orthogonal Lie algebra

$$L(G) = so_n(q) = \{A \in \mathrm{End}(V) : (vA, w) + (v, wA) = 0\ \forall v, w \in V\}.$$

Odd Blocks $V_\beta(2m+1)$

Let α be a fixed non-square in $\mathbb{F}_q^*$, and let $\beta \in \{1,\alpha\}$. Let V_{2m+1} be a $(2m+1)$-dimensional vector space over $\mathbb{F}_q$, with basis $v_{-2m}, v_{-(2m-2)}, \ldots, v_{2m}$. Define an orthogonal form on V_{2m+1} by setting $(v_i, v_{-i}) = (-1)^{i/2}$ for all $i \neq 0$, $(v_0, v_0) = \beta$, and all other values $(v_i, v_j) = 0$. Define $e_\beta \in \mathrm{End}(V_{2m+1})$ by

$$\begin{aligned} e_\beta : v_i &\mapsto v_{i+2} \ (i \neq 0, 2m),\\ v_0 &\mapsto \beta v_2,\\ v_{2m} &\mapsto 0. \end{aligned}$$

Now e_β is a nilpotent element of $so_{2m+1}(q)$. As in Sect. 4.1.1, we apply the Cayley map and set

$$V_\beta(2m+1) = (1-e_\beta)(1+e_\beta)^{-1} \in G = \mathrm{O}(V_{2m+1}).$$

Observe that $V_\beta(2m+1)$ is a single Jordan block. Its matrix relative to the given basis is the $(2m+1)$-dimensional version of the matrix given for $V_\beta(2m)$ in the symplectic case. It is conjugate to the element labelled $V_{2\beta}(2m+1)$ in [14, §2.5].

Define a 1-dimensional torus $T_1 = \{T_1(c) : c \in \bar{\mathbb{F}}_q^*\} \leq \mathrm{O}(\bar{V}_{2m+1})$, where $\bar{V}_{2m+1} = V_{2m+1} \otimes \bar{\mathbb{F}}_q$, by the action

$$T_1(c) : v_i \mapsto c^i v_i \text{ for all } i.$$

Note that $e_\beta T_1(c) = c^2 e_\beta$.

Even Blocks $W(2m)$

Let V_{4m} be a $4m$-dimensional space over $\mathbb{F}_q$ with basis

$$w_{-(2m-1)}, x_{-(2m-1)}, w_{-(2m-3)}, x_{-(2m-3)}, \ldots, w_{2m-1}, x_{2m-1},$$

and orthogonal form defined by $(w_i, x_{-i}) = (-1)^{\lfloor i/2 \rfloor}$ for all i and all other values (w_i, w_j), (w_i, x_j), (x_i, x_j) equal 0. Define $e \in \mathrm{End}(V_{4m+2})$ by

$$\begin{aligned} e : w_i &\mapsto w_{i+2} \ (i < 2m-1),\\ x_i &\mapsto x_{i+2} \ (i < 2m-1),\\ w_{2m-1} &\mapsto 0,\\ x_{2m-1} &\mapsto 0. \end{aligned}$$

Now e is a nilpotent element of $so_{4m}(q)$. Set

$$W(2m) = (1-e)(1+e)^{-1} \in G = \mathrm{O}(V_{4m}).$$

Observe that $W(2m)$ has Jordan form $J_{2m}^{[2]}$. Its matrix relative to the given basis is

$$\begin{pmatrix} I & -2I & 2I & \cdots & -2I \\ & I & -2I & \cdots & 2I \\ & & & \ddots & \\ & & & & I \end{pmatrix}$$

where I denotes the 2×2 identity matrix. Define a 1-dimensional torus $T_1 \leq \mathrm{O}(\bar{V}_{4m})$ by the action

$$T_1(c) : w_i \mapsto c^i w_i,\ x_i \mapsto c^i x_i \quad \text{for all } i.$$

Class Representatives

For a given Jordan form (4.1.4), the conjugacy class representatives in $\mathrm{O}(V)$ with this Jordan form are

$$\bigoplus_{i=1}^{r} (V_{\beta_i}(2k_i+1) \oplus V_1(2k_i+1)^{[a_i-1]}) \oplus \bigoplus_{i=1}^{s} W(2l_i)^{[b_i]}, \tag{4.1.5}$$

where $V = \bigoplus_{i=1}^{r} V_{2k_i+1}^{[a_i]} \oplus \bigoplus_{i=1}^{s} V_{4l_i}^{[b_i]}$, and each $\beta_i \in \{1, \alpha\}$ (and the k_i are distinct, as are the l_i).

Let u be the element defined in (4.1.5). Then u fixes an orthogonal form of discriminant $D = \prod_{i=1}^{r} \beta_i (-1)^{k_i a_i}$ (set $D = 1$ if $r = 0$). If $n = \dim V = 2m$ is even, then $u \in \mathrm{O}_{2m}^{\epsilon}(q)$, where $\epsilon = +$ if and only if $(-1)^m D$ is a square in $\mathbb{F}_q^*$ (see [21, Prop. 2.5.10]). For the various choices of β_i, these representatives fall into 2^{r-1} conjugacy classes in each of $\mathrm{O}_{2m}^{+}(q)$ and $\mathrm{O}_{2m}^{-}(q)$, except that for $r = 0$, there is one class in $\mathrm{O}_{2m}^{+}(q)$ and none in $\mathrm{O}_{2m}^{-}(q)$. If n is odd, then half of the representatives (2^{r-1} of them) lie in a given $\mathrm{O}_n(q)$ fixing a form of square discriminant.

Define T to be the 1-dimensional torus acting as $T_1(c)$ on each subspace $\bar{V}_{2k_i}$ and $\bar{V}_{4l_i+2}$. To define corresponding matrix representatives, choose an ordered basis of V according to the T-weights. Namely, list the spaces V_{2k_j+1}, V_{4l_j} in decreasing order of Jordan block sizes. For the ith space, if it is $V_\beta(2k+1)$, then label its basis as

$$v_{-(2k)}^{(i)}, v_{-(2k-2)}^{(i)}, \ldots, v_{2k}^{(i)},$$

and if it is $W(2l)$, then label its basis as

$$w_{-(2l-1)}^{(i)}, x_{-(2l-1)}^{(i)}, w_{-(2l-3)}^{(i)}, x_{-(2l-3)}^{(i)}, \ldots, w_{2l-1}^{(i)}, x_{2l-1}^{(i)}.$$

Now take a basis of V consisting of these vectors $v_j^{(i)}, w_j^{(i)}, x_j^{(i)}$ in increasing order lexicographically with respect to the pairs (j, i). Choose the matrix representative corresponding to the linear map (4.1.5) to be its matrix with respect to this ordered basis.

Let P be the parabolic subgroup stabilizing the flag of V defined by sums of T-weight spaces for decreasing weights; so $P = QL$, with unipotent radical Q and Levi subgroup $L = C_G(T)$. As in the previous section, we have the following result.

Theorem 4.1.3 *Let $G = \mathrm{O}_n^\epsilon(q)$ with q odd, and let $u \in G$ be the unipotent matrix representative defined above for the linear map*

$$\bigoplus_{i=1}^{r}(V_{\beta_i}(2k_i+1) \oplus V_1(2k_i+1)^{[a_i-1]}) \oplus \bigoplus_{i=1}^{s} W(2l_i)^{[b_i]}.$$

Let $P = QL$ be the corresponding parabolic subgroup.

(i) $C_G(u) = C_Q(u)C_L(u)$;
(ii) $C_L(u) \cong \prod_{i=1}^{s} \mathrm{Sp}_{2b_i}(q) \times \prod_{i=1}^{r} \mathrm{O}_{a_i}^{\epsilon_i}(q)$;
(iii) *rewrite the Jordan form of u as $\sum J_i^{r_i}$; then $|C_Q(u)| = q^m$ where*

$$m = \frac{1}{2}\sum_i (i-1)r_i^2 + \sum_{i<j} i r_i r_j - \frac{1}{2}\sum_{i\ even} r_i.$$

In part (ii), the sign $\epsilon_i = \pm$ is the sign of the orthogonal form with Gram matrix $\mathrm{diag}(\beta_i, 1, \ldots, 1)$. In Sect. 4.2.2 we justify this, and show how to construct the subgroups $C_Q(u)$ and $C_L(u)$ of the centralizer.

Theorem 4.1.3 gives the unipotent class representatives u in the orthogonal group $G = \mathrm{O}_n^\epsilon(q)$. We conclude by describing how these classes split within the special orthogonal group $\mathrm{SO}_n^\epsilon(q)$ and its subgroup $\Omega_n^\epsilon(q)$ (see [14, Prop. 2.4]). First, a class u^G splits into two $\mathrm{SO}_n^\epsilon(q)$-classes if and only if $\epsilon = +$ and $r = 0$; in this case, representatives are u and u^t for t a reflection in G. Secondly, an $\mathrm{SO}_n^\epsilon(q)$-class with representative u (as in (4.1.5)) splits into two $\Omega_n^\epsilon(q)$-classes if and only if either $r = 0$, or $r \geq 1$ and the following hold:

(a) $a_i = 1$ for all i, and
(b) $\beta_i \equiv (-1)^{k_1+k_i}\beta_1 \bmod (\mathbb{F}_q^*)^2$ for all i.

In case of splitting, representatives of the $\Omega_n^\epsilon(q)$-classes are u and u^s, where $s \in \mathrm{SO}_n^\epsilon(q) \setminus \Omega_n^\epsilon(q)$.

4.1.3 Unitary Groups

Let $\mathrm{GU}_n(q)$ fix a unitary form on an n-dimensional vector space V over $\mathbb{F}_{q^2}$, and let $\alpha \mapsto \bar{\alpha}$ denote the involutory automorphism of $\mathbb{F}_{q^2}$. Here we take the unipotent class representatives directly from [14, §2.3]. To define these, we need to describe unitary Jordan blocks of even and odd sizes. Fix $\beta, \gamma \in \mathbb{F}_{q^2}^*$ with $\beta + \bar{\beta} = 0$, $\gamma + \bar{\gamma} = -1$.

Even Blocks $V(2m)$

Let V_{2m} be a $2m$-dimensional space over $\mathbb{F}_{q^2}$ with basis $v_{-(2m-1)}, v_{-(2m-3)}, \ldots, v_{2m-1}$, and unitary form defined by $(v_{-i}, v_i) = 1$ for all i, and all other $(v_i, v_j) = 0$. Define a unipotent element of the unitary group $\mathrm{SU}(V_{2m})$ as follows:

$$
\begin{aligned}
V(2m) : v_{-i} &\mapsto v_{-i} + v_{-i+2} + \cdots + v_{-1} + \beta v_1 \quad (i = 1, 3, \ldots, 2m-1)\\
v_j &\mapsto v_j - v_{j+2} \quad (j = 1, 3, \ldots, 2m-1).
\end{aligned}
$$

Odd Blocks $V(2m + 1)$

Let V_{2m+1} be a $(2m+1)$-dimensional space over $\mathbb{F}_{q^2}$ with basis $v_{-2m}, v_{-(2m-2)}, \ldots, v_{2m}$, and unitary form defined by $(v_{-i}, v_i) = 1$ for all i, and all other $(v_i, v_j) = 0$. Define a unipotent element of the unitary group $\mathrm{SU}(V_{2m+1})$ as follows:

$$
\begin{aligned}
V(2m+1) : v_{-i} &\mapsto v_{-i} + v_{-i+2} + \cdots + v_0 + \gamma v_2 \quad (i = 2, 4, \ldots, 2m)\\
v_j &\mapsto v_j - v_{j+2} \quad (j = 0, 2, \ldots, 2m).
\end{aligned}
$$

Class Representatives

Denote by $\bigoplus_{i=1}^{s} V(m_i)^{[r_i]}$ the orthogonal direct sum of copies of the linear maps defined above acting on $V = \bigoplus_{i=1}^{s} V_{m_i}^{[r_i]}$, where $m_1 > m_2 > \cdots > m_s$. This has Jordan form $\bigoplus_{i=1}^{s} J_{m_i}^{[r_i]}$, and there is a unique conjugacy class in $\mathrm{GU}(V)$ having this Jordan form. The number of conjugacy classes in $\mathrm{SU}(V)$ with this Jordan form is $\gcd(t, q+1)$, where $t = \gcd(m_1, \ldots, m_s)$ (see [14, Prop. 2.2]). Representatives can be found by taking conjugates of $\bigoplus_{i=1}^{s} V(m_i)^{[r_i]}$ by matrices $d(\alpha) \in \mathrm{GU}(V)$ of determinant α, where $Z = \{\mu \in \mathbb{F}_{q^2} : \mu\bar{\mu} = 1\}$ and α ranges over representatives of Z/Z^t. For example, we can take $d(\alpha)$ to act on the basis of one of the blocks as $\mathrm{diag}(\lambda, 1, \ldots, 1, \bar{\lambda}^{-1})$, where $\alpha = \lambda\bar{\lambda}^{-1}$.

To define matrix representatives, list the spaces V_{m_j} in decreasing order of sizes, and for the ith space, if it is $V(m)$, label its basis as $v_{-(m-1)}^{(i)}, \ldots, v_{m-1}^{(i)}$. Now take a basis of V consisting of these vectors $v_j^{(i)}$ for all i, j, in increasing order lexicographically with respect to the pairs (j, i). Choose the matrix representative u

corresponding to $\bigoplus_{i=1}^{s} V(m_i)^{[r_i]}$ to be the matrix of the linear map with respect to this ordered basis.

For each i, define $V_{(i)}$ to be the span of the basis vectors with subscript i, and call this the weight space for i. As in previous sections, let P be the parabolic subgroup stabilizing the flag of V defined by sums of weight spaces for decreasing weights; so $P = QL$, where Q is the unipotent radical and L a Levi subgroup. The next result follows from [22, Lemma 2.29 and Props. 3.7, 3.8].

Theorem 4.1.4 *Let $G = \mathrm{GU}_n(q)$, let $u \in G$ be the unipotent matrix defined above with Jordan form $\bigoplus_{i=1}^{s} J_{m_i}^{[r_i]}$, and let $P = QL$ be the corresponding parabolic subgroup.*

(i) $C_G(u) = C_Q(u)C_L(u)$;
(ii) $C_L(u) \cong \prod_{i=1}^{s} \mathrm{GU}_{r_i}(q)$;
(iii) $|C_Q(u)| = q^m$, *where* $m = \sum_i (i-1)r_i^2 + 2\sum_{i<j} i r_i r_j$.

4.2 Centralizers of Class Representatives

Having described the unipotent class representatives, we now show how to construct their centralizers, whose structures are given in Theorems 4.1.2, 4.1.3 and 4.1.4. In these results, for each class representative u of an isometry group G, we defined a parabolic subgroup $P = QL$ such that $C_G(u) = C_Q(u)C_L(u)$ and gave the structure of $C_L(u)$ and the order of $C_Q(u)$.

We construct $C_Q(u)$ and $C_L(u)$ in the following steps:

(a) construct a generating set for Q;
(b) construct a *power-conjugate presentation* for Q and use it to compute generators for $C_Q(u)$;
(c) construct a generating set for $C_L(u)$.

We comment briefly on Step (b). Recall that Q is a unipotent matrix group. A group of prime-power order can be described by a power-conjugate presentation which reflects a central series; various problems—including deciding conjugation and constructing centralizers—can be solved easily in practice by exploiting this presentation. It is a routine computational exercise to construct an isomorphic copy $\bar{Q}$ of Q described by such a presentation; the centralizer of an element of Q can then be readily constructed in $\bar{Q}$, and so in Q. For related details, see [17, Chap. 9].

In the rest of the section, we describe Steps (a) and (c) for each type of classical group.

4.2.1 Symplectic Groups

Let $G = \mathrm{Sp}(V) \cong \mathrm{Sp}_{2n}(q)$, and let u be the matrix representative of the element

$$\bigoplus_{i=1}^{r}(V_{\beta_i}(2k_i) \oplus V_1(2k_i)^{[a_i-1]}) \oplus \bigoplus_{i=1}^{s} W(2l_i+1)^{[b_i]}, \tag{4.2.1}$$

of $\mathrm{Sp}(V)$ as in (4.1.3), relative to the ordered basis of vectors $v_j^{(i)}, w_j^{(i)}, x_j^{(i)}$ described in Sect. 4.1.1. Let $P = QL$ be the parabolic subgroup defined there.

Generators of Q

We define two collections of generators for Q.

(1) For all i, j with $i < 0$, $i \leq j < -i$, all k, l, and every $\lambda \in \mathbb{F}_q$ define the following generators which fix all basis vectors other than those listed. Set $\epsilon = (-1)^{\lfloor i/2 \rfloor + \lfloor j/2 \rfloor}$.

$$\begin{aligned} x_{1ijkl}(\lambda) : v_i^{(k)} &\mapsto v_i^{(k)} + \lambda v_{-j}^{(l)}, \\ v_j^{(l)} &\mapsto v_j^{(l)} + \epsilon\lambda v_{-i}^{(k)} \end{aligned}$$

$$\begin{aligned} x_{2ijkl}(\lambda) : v_i^{(k)} &\mapsto v_i^{(k)} + \lambda x_{-j}^{(l)}, \\ w_j^{(l)} &\mapsto w_j^{(l)} + \epsilon\lambda v_{-i}^{(k)} \end{aligned}$$

$$\begin{aligned} x_{3ijkl}(\lambda) : v_j^{(k)} &\mapsto v_j^{(k)} + \lambda x_{-i}^{(l)}, \\ w_i^{(l)} &\mapsto w_i^{(l)} + \epsilon\lambda v_{-j}^{(k)} \end{aligned}$$

$$\begin{aligned} x_{4ijkl}(\lambda) : v_i^{(k)} &\mapsto v_i^{(k)} + \lambda w_{-j}^{(l)}, \\ x_j^{(l)} &\mapsto x_j^{(l)} - \epsilon\lambda v_{-i}^{(k)} \end{aligned}$$

$$x_{5ijkl}(\lambda) : v_j^{(k)} \mapsto v_j^{(k)} + \lambda w_{-i}^{(l)},$$
$$x_i^{(l)} \mapsto x_i^{(l)} - \epsilon\lambda v_{-j}^{(k)}$$

$$x_{6ijkl}(\lambda) : w_i^{(k)} \mapsto w_i^{(k)} + \lambda x_{-j}^{(l)},$$
$$w_j^{(l)} \mapsto w_j^{(l)} + \epsilon\lambda x_{-i}^{(k)}$$

$$x_{7ijkl}(\lambda) : x_i^{(k)} \mapsto x_i^{(k)} + \lambda w_{-j}^{(l)},$$
$$x_j^{(l)} \mapsto x_j^{(l)} + \epsilon\lambda w_{-i}^{(k)}$$

$$x_{8ijkl}(\lambda) : w_i^{(k)} \mapsto w_i^{(k)} + \lambda w_{-j}^{(l)},$$
$$x_j^{(l)} \mapsto x_j^{(l)} - \epsilon\lambda x_{-i}^{(k)}$$

$$x_{9ijkl}(\lambda) : x_i^{(k)} \mapsto x_i^{(k)} + \lambda x_{-j}^{(l)},$$
$$w_j^{(l)} \mapsto w_j^{(l)} - \epsilon\lambda w_{-i}^{(k)}.$$

(2) For all $i > 0$, all k, and every $\lambda \in \mathbb{F}_q$:

$$y_{1ik}(\lambda) : v_{-i}^{(k)} \mapsto v_{-i}^{(k)} + \lambda v_i^{(k)},$$

$$y_{2ik}(\lambda) : w_{-i}^{(k)} \mapsto w_{-i}^{(k)} + \lambda x_i^{(k)},$$

$$y_{3ik}(\lambda) : x_{-i}^{(k)} \mapsto x_{-i}^{(k)} + \lambda w_i^{(k)}.$$

Define $X_{cijkl} = \{x_{cijkl}(\lambda) : \lambda \in \mathbb{F}_q\} \cong \mathbb{F}_q^+$ for each of the generators listed; similarly define Y_{mik}. Of course, to construct each subgroup we need only write down generators for those λ in a basis of $\mathbb{F}_q$ over $\mathbb{F}_p$.

To check that these subgroups generate Q, observe that each lies in Q, and the total number of subgroups is m, where $|Q| = q^m$.

Example 4.2.1 We return to Example 4.1.1. Here

$$u = V_\alpha(4) \oplus W(3) \oplus V_1(2) \in \mathrm{Sp}_{12}(q).$$

The parabolic subgroup $P = QL$ has $L = \mathrm{GL}_1(q) \times \mathrm{GL}_2(q) \times \mathrm{GL}_2(q) \times \mathrm{Sp}_2(q)$, and $|Q| = q^{33}$. We record the number of subgroups X_{1ijkl}, ..., X_{9ijkl}, Y_{1ik}, Y_{2ik}, and Y_{3ik}:

Subgroups	Number
X_{1ijkl}	4
X_{2ijkl}	5
X_{3ijkl}	4
X_{4ijkl}	5
X_{5ijkl}	4
X_{6ijkl}	1
X_{7ijkl}	1
X_{8ijkl}	2
X_{9ijkl}	2
Y_{1ik}	3
Y_{2ik}	1
Y_{3ik}	1

The total number of subgroups is 33, as claimed.

Generators of $C_L(u)$

For $u \in \mathrm{Sp}_{2n}(q)$ as in (4.2.1), Theorem 4.1.2 gives

$$C_L(u) \cong \prod_{i=1}^{s} \mathrm{Sp}_{2b_i}(q) \times \prod_{i=1}^{r} \mathrm{O}_{a_i}^{\epsilon_i}(q).$$

Each factor $\mathrm{Sp}_{2b_i}(q)$ acts on the summand $W(2l_i + 1)^{[b_i]}$, fixing the orthogonal complement; and each factor $\mathrm{O}_{a_i}^{\epsilon_i}(q)$ acts on the summand $V_{\beta_i}(2k_i) \oplus V_1(2k_i)^{[a_i-1]}$. So it suffices to focus on these summands. For notational convenience, we denote them by $W(2l + 1)^{[b]}$ and $V_\beta(2k) \oplus V_1(2k)^{[a-1]}$.

Summand $W(2l + 1)^{[b]}$ Let $u = W(2l + 1)^{[b]}$. We need to find generators for $C_L(u) = \mathrm{Sp}_{2b}(q)$. The ordered basis of $W(2l + 1)^{[b]}$ is

$$w_{-2l}^{(1)}, x_{-2l}^{(1)}, \ldots, w_{-2l}^{(b)}, x_{-2l}^{(b)}, \ldots, w_{2l}^{(1)}, x_{2l}^{(1)}, \ldots, w_{2l}^{(b)}, x_{2l}^{(b)}.$$

With respect to this basis, the symplectic form has matrix

$$\begin{pmatrix} & & & K \\ & & -K & \\ & \iddots & & \\ K & & & \end{pmatrix}$$

where $K = J^{[b]}$ (block diagonal sum of b copies of $J = \begin{pmatrix} & -1 \\ 1 & \end{pmatrix}$). With respect to the above basis, the factor $L_0 \cong \mathrm{Sp}_{2b}(q)$ of $C_L(u)$ is

$$L_0 = \left\{ \begin{pmatrix} A & & & \\ & A & & \\ & & \ddots & \\ & & & A \end{pmatrix} : A \in \mathrm{Sp}_{2b}(q) \right\},$$

where $\mathrm{Sp}_{2b}(q)$ stabilizes the symplectic form with matrix K. We choose standard generators for this group (as mentioned at the end of Sect. 3.1), and hence obtain generators for L_0.

Summand $V_\beta(2k) \oplus V_1(2k)^{[a-1]}$ Here the basis is

$$v_{-(2k-1)}^{(1)}, \ldots, v_{-(2k-1)}^{(a)}, \ldots, v_{2k-1}^{(1)}, \ldots, v_{2k-1}^{(a)}.$$

Change the basis as follows: $v_1^{(1)} \mapsto \beta v_1^{(1)}, \ldots, v_{2k-1}^{(1)} \mapsto \beta v_{2k-1}^{(1)}$. With respect to this basis, the factor $L_0 \cong \mathrm{O}_a^\epsilon(q)$ is

$$L_0 = \left\{ \begin{pmatrix} A & & & \\ & A & & \\ & & \ddots & \\ & & & A \end{pmatrix} : A \in \mathrm{O}_a^\epsilon(q) \right\},$$

and $\mathrm{O}_a^\epsilon(q)$ stabilizes the orthogonal form with matrix $\mathrm{diag}(\beta, 1, \ldots, 1)$. Now choose standard generators for this group, and then change back to the original basis to obtain the required factor $\mathrm{O}_a^\epsilon(q)$ of $C_L(u)$.

4.2.2 Orthogonal Groups

This is similar to the previous section. Let $G = \mathrm{O}(V) \cong \mathrm{O}^{\epsilon}_{2n}(q)$, and let u be the matrix representative of the element

$$\bigoplus_{i=1}^{r}(V_{\beta_i}(2k_i+1) \oplus V_1(2k_i+1)^{[a_i-1]}) \oplus \bigoplus_{i=1}^{s} W(2l_i)^{[b_i]}, \tag{4.2.2}$$

of $\mathrm{O}(V)$ as in (4.1.5), relative to the ordered basis of vectors $v_j^{(i)}$, $w_j^{(i)}$, $x_j^{(i)}$ described in Sect. 4.1.2. Let $P = QL$ be the parabolic subgroup defined there.

Generators of Q

We define two collections of generators for Q.

(1) For all i, j with $i < 0$, $i \le j < -i$ and $j \ne 0$, all k, l, and every $\lambda \in \mathbb{F}_q$ define the following generators which fix all basis vectors other than those listed. Set $\epsilon_1 = (-1)^{\lfloor i/2 \rfloor + \lfloor j/2 \rfloor}$, $\epsilon_2 = (-1)^{\lfloor i/2 \rfloor + \lfloor -j/2 \rfloor}$, and $\epsilon_3 = (-1)^{\lfloor -i/2 \rfloor + \lfloor -j/2 \rfloor}$.

$$\begin{aligned} x_{1ijkl}(\lambda) : v_i^{(k)} &\mapsto v_i^{(k)} + \lambda v_{-j}^{(l)}, \\ v_j^{(l)} &\mapsto v_j^{(l)} - \epsilon_1 \lambda v_{-i}^{(k)} \end{aligned}$$

$$\begin{aligned} x_{2ijkl}(\lambda) : v_i^{(k)} &\mapsto v_i^{(k)} + \lambda x_{-j}^{(l)}, \\ w_j^{(l)} &\mapsto w_j^{(l)} - \epsilon_1 \lambda v_{-i}^{(k)} \end{aligned}$$

$$\begin{aligned} x_{3ijkl}(\lambda) : v_j^{(k)} &\mapsto v_j^{(k)} + \lambda x_{-i}^{(l)}, \\ w_i^{(l)} &\mapsto w_i^{(l)} - \epsilon_1 \lambda v_{-j}^{(k)} \end{aligned}$$

$$\begin{aligned} x_{4ijkl}(\lambda) : v_i^{(k)} &\mapsto v_i^{(k)} + \lambda w_{-j}^{(l)}, \\ x_j^{(l)} &\mapsto x_j^{(l)} - \epsilon_1 \lambda v_{-i}^{(k)} \end{aligned}$$

$$\begin{aligned} x_{5ijkl}(\lambda) : v_j^{(k)} &\mapsto v_j^{(k)} + \lambda w_{-i}^{(l)}, \\ x_i^{(l)} &\mapsto x_i^{(l)} - \epsilon_1 \lambda v_{-j}^{(k)} \end{aligned}$$

$$x_{6ijkl}(\lambda): w_i^{(k)} \mapsto w_i^{(k)} + \lambda x_{-j}^{(l)},$$
$$w_j^{(l)} \mapsto w_j^{(l)} - \epsilon_1 \lambda x_{-i}^{(k)}$$

$$x_{7ijkl}(\lambda): x_i^{(k)} \mapsto x_i^{(k)} + \lambda w_{-j}^{(l)},$$
$$x_j^{(l)} \mapsto x_j^{(l)} - \epsilon_3 \lambda w_{-i}^{(k)}$$

$$x_{8ijkl}(\lambda): w_i^{(k)} \mapsto w_i^{(k)} + \lambda w_{-j}^{(l)},$$
$$x_j^{(l)} \mapsto x_j^{(l)} - \epsilon_2 \lambda x_{-i}^{(k)}$$

$$x_{9ijkl}(\lambda): x_i^{(k)} \mapsto x_i^{(k)} + \lambda x_{-j}^{(l)},$$
$$w_j^{(l)} \mapsto w_j^{(l)} - \epsilon_2 \lambda w_{-i}^{(k)}.$$

(2) For all $i > 0$, all k, l, and every $\lambda \in \mathbb{F}_q$, define the following generators which fix all basis vectors other than those listed. Set $\mu_\epsilon = -(-1)^{\lfloor \epsilon i/2 \rfloor} \beta\lambda$, and $\alpha_\epsilon = -\frac{1}{2}(-1)^{\lfloor \epsilon i/2 \rfloor} \beta\lambda^2$, where $\epsilon = \pm$.

$$\begin{aligned} y_{1ikl}(\lambda): v_{-i}^{(k)} &\mapsto v_{-i}^{(k)} + \lambda v_0^{(l)} + \alpha_+ v_i^{(k)}, \\ v_0^{(l)} &\mapsto v_0^{(l)} + \mu_+ v_i^{(k)} \end{aligned}$$

$$\begin{aligned} y_{2ikl}(\lambda): w_{-i}^{(k)} &\mapsto w_{-i}^{(k)} + \lambda v_0^{(l)} + \alpha_- x_i^{(k)}, \\ v_0^{(l)} &\mapsto v_0^{(l)} + \mu_- x_i^{(k)} \end{aligned}$$

$$\begin{aligned} y_{3ikl}(\lambda): x_{-i}^{(k)} &\mapsto x_{-i}^{(k)} + \lambda v_0^{(l)} + \alpha_+ w_i^{(k)}, \\ v_0^{(l)} &\mapsto v_0^{(l)} + \mu_+ w_i^{(k)}. \end{aligned}$$

As in the previous section, define $X_{cijkl} = \{x_{cijkl}(\lambda) : \lambda \in \mathbb{F}_q\} \cong \mathbb{F}_q^+$ for each of the generators listed; similarly define Y_{mikl}. Each of these subgroups lies in Q, and the total number is m, where $|Q| = q^m$.

Generators of $C_L(u)$

For $u \in \mathrm{O}_n^\epsilon(q)$ as in (4.2.2), Theorem 4.1.3 gives

$$C_L(u) \cong \prod_{i=1}^{s} \mathrm{Sp}_{2b_i}(q) \times \prod_{i=1}^{r} \mathrm{O}_{a_i}^{\epsilon_i}(q).$$

Each factor $\mathrm{Sp}_{2b_i}(q)$ acts on the summand $W(2l_i)^{[b_i]}$, fixing the orthogonal complement; and each factor $\mathrm{O}_{a_i}^{\epsilon_i}(q)$ acts on the summand $V_{\beta_i}(2k_i+1) \oplus V_1(2k_i+1)^{[a_i-1]}$. So it suffices to focus on these summands. For notational convenience, we denote them by $W(2l)^{[b]}$ and $V_\beta(2k+1) \oplus V_1(2k+1)^{[a-1]}$.

Summand $W(2l)^{[b]}$ Let $u = W(2l)^{[b]}$. As in the previous section, with respect to the basis $w_i^{(j)}, x_i^{(j)}$, the factor $L_0 \cong \mathrm{Sp}_{2b}(q)$ of $C_L(u)$ is

$$L_0 = \left\{ \begin{pmatrix} A & & & \\ & A & & \\ & & \ddots & \\ & & & A \end{pmatrix} : A \in \mathrm{Sp}_{2b}(q) \right\},$$

where $\mathrm{Sp}_{2b}(q)$ stabilizes the symplectic form with matrix $K = J^{[b]}$, as before. We choose standard generators for this group, and hence obtain generators for L_0.

Summand $V_\beta(2k+1) \oplus V_1(2k+1)^{[a-1]}$ Here the basis is

$$v_{-(2k)}^{(1)}, \dots, v_{-(2k)}^{(a)}, \dots, v_{2k}^{(1)}, \dots, v_{2k}^{(a)}.$$

Change the basis as follows: $v_2^{(1)} \mapsto \beta v_2^{(1)}, \dots, v_{2k}^{(1)} \mapsto \beta v_{2k}^{(1)}$. Let $\mathrm{O}_a^\delta(q)$ be the orthogonal group stabilizing the orthogonal form with matrix $\mathrm{diag}(\beta, 1, \dots, 1)$. With respect to the above basis, the factor $L_0 \cong \mathrm{O}_a^\delta(q)$ is

$$L_0 = \left\{ \begin{pmatrix} A & & & \\ & A & & \\ & & \ddots & \\ & & & A \end{pmatrix} : A \in \mathrm{O}_a^\delta(q) \right\}.$$

Now choose standard generators for this group, and then change back to the original basis to obtain the required factor $\mathrm{O}_a^\delta(q)$ of $C_L(u)$.

We can use Schreier's Lemma and the consequent algorithm [17, Chap. 2] to write down the kernel of an explicit homomorphism from a group into a cyclic group. Let $\mathcal{S} = \mathrm{SO}^\epsilon(V)$ and $\Omega = \Omega^\epsilon(V)$. Both $C_G(u)/C_{\mathcal{S}}(u)$ and $C_{\mathcal{S}}(u)/C_\Omega(u)$ are cyclic, and are the images of the determinant and spinor norm homomorphisms respectively. We apply this algorithm to construct generating sets for $C_{\mathcal{S}}(u)$ and $C_\Omega(u)$.

4.2.3 Unitary Groups

Let $G = \mathrm{GU}(V) \cong \mathrm{GU}_n(q)$, and let u be the matrix representative of the element $\bigoplus_{i=1}^{s} V(m_i)^{[r_i]}$ relative to the ordered basis of vectors $v_j^{(i)}$ described in Sect. 4.1.3. Let $P = QL$ be the parabolic subgroup defined there.

Generators of Q

(1) For all $i, j > 0$ with $-i \le j < i$ and $j \neq 0$, all k, l, and every $\lambda \in \mathbb{F}_{q^2}$ define

$$\begin{aligned} x_{1ijkl}(\lambda) : v_{-i}^{(k)} &\mapsto v_{-i}^{(k)} + \lambda v_{-j}^{(l)}, \\ v_j^{(l)} &\mapsto v_j^{(l)} - \bar{\lambda} v_i^{(k)}. \end{aligned}$$

(2) If there are no weight 0 vectors in the basis (no vectors $v_0^{(i)}$), then for all $i < 0$, all k, and every $\alpha \in \mathbb{F}_{q^2}$ such that $\alpha + \bar{\alpha} = 0$, define

$$x_{2ikl}(\lambda) : v_{-i}^{(k)} \mapsto v_{-i}^{(k)} + \alpha v_i^{(k)}.$$

(3) If there are weight 0 vectors in the basis, then for all $i < 0$, all k, l, and every $\lambda, \alpha \in \mathbb{F}_{q^2}$ such that $\lambda\bar{\lambda} + \alpha + \bar{\alpha} = 0$, define

$$\begin{aligned} x_{3ikl}(\lambda, \alpha) : v_{-i}^{(k)} &\mapsto v_{-i}^{(k)} + \lambda v_0^{(l)} + \alpha v_i^{(k)}, \\ v_0^{(l)} &\mapsto v_0^{(l)} - \bar{\lambda} v_i^{(k)}. \end{aligned}$$

As in previous sections, these elements generate Q.

Generators of $C_L(u)$

Let $u = \bigoplus_{i=1}^{s} V(m_i)^{[r_i]} \in \mathrm{GU}_n(q)$. Theorem 4.1.4 gives

$$C_L(u) \cong \prod_{i=1}^{s} \mathrm{GU}_{r_i}(q).$$

We work in a summand $V(m_i)^{[r_i]}$, which we write as $V(m)^{[r]}$ for simplicity, with basis

$$v_{-(m-1)}^{(1)}, \ldots, v_{-(m-1)}^{(r)}, \ldots, v_{m-1}^{(1)}, \ldots, v_{m-1}^{(r)},$$

where the unitary form is as in Sect. 4.1.3. Now $C_L(u)$ is

$$\left\{ \begin{pmatrix} A & & & \\ & A & & \\ & & \ddots & \\ & & & A \end{pmatrix} : A \in \mathrm{GU}_r(q) \right\},$$

where $\mathrm{GU}_r(q)$ stabilizes the unitary form with matrix I_r. We use standard generators for $\mathrm{GU}_r(q)$ to construct $C_L(u)$.

To obtain $C_{\mathrm{SU}_r(q)}(u)$ we must construct $C := C_L(u) \cap \mathrm{SU}_r(q)$. Using Schreier's algorithm, we construct C as the kernel of the determinant homomorphism from $C_L(u)$ into $\mathbb{F}_{q^2}^*$.

4.3 The Conjugacy Problem

In this section we solve the conjugacy problem: for a unipotent element g of a classical group G, identify the class representative defined in Sect. 4.1 that is conjugate to g. By two applications of our solution, we then solve the general problem: given unipotent $g, h \in G$, is g conjugate to h?

4.3.1 Symplectic Groups

Let $G = \mathrm{Sp}(V) \cong \mathrm{Sp}_{2n}(q)$ with q odd, preserving a symplectic form (,), and let $g \in G$ be unipotent. We want to compute the representative

$$\bigoplus_{i=1}^{r} (V_{\beta_i}(2k_i) \oplus V_1(2k_i)^{[a_i-1]}) \oplus \bigoplus_{i=1}^{s} W(2l_i+1)^{[b_i]}, \tag{4.3.1}$$

as defined in Sect. 4.1.1, such that g is G-conjugate to this representative. The values $k_1, \ldots, k_r$ and $l_1, \ldots, l_s$ are determined by the Jordan form of g. It remains to determine the values of the parameters $\beta_i \in \{1, \alpha\}$ in (4.3.1).

Our algorithm to do this uses the following two lemmas. The *radical* of $F \leq V$ is $\mathrm{Rad}(F) = F \cap F^{\perp}$.

Lemma 4.3.1 *Let* $u = V_\beta(2k) \in \mathrm{Sp}(V)$ *with* $V = V_{2k}$, *as defined in Sect. 4.1.1. Let* $0 \leq r < k$, *and* $F = V(1-u)^r$.

(i) $u^{F/\mathrm{Rad}(F)}$, *the element induced by the action of* u *on the symplectic space* $F/\mathrm{Rad}(F)$, *is* $V_\beta(2k-2r)$.
(ii) *If* $r = 2$, *then* $F^{\perp}$ *is totally isotropic.*

Proof

(i) Now V_{2k} has basis $v_{-(2k-1)}, v_{-(2k-3)}, \ldots, v_{2k-1}$, and symplectic form given by $(v_i, v_{-i}) = (-1)^{\lfloor i/2 \rfloor}$ for all i, and $u = (1 - e_\beta(2k))(1 + e_\beta(2k))^{-1}$ where $e_\beta(2k)$ is the nilpotent linear map defined by

$$\begin{aligned} e_\beta(2k) : v_i &\mapsto v_{i+2} \;\; (i < 2m-1,\, i \neq -1), \\ v_{-1} &\mapsto \beta v_1, \\ v_{2m-1} &\mapsto 0. \end{aligned}$$

Since $e_\beta(2k) = (1+u)^{-1}(1-u)$,

$$F = V(1-u)^r = Ve_\beta(2k)^r = \langle v_{-(2k-2r-1)}, \ldots, v_{2k-1} \rangle.$$

Then $\mathrm{Rad}(F) = \langle v_{2k-2r+1}, \ldots, v_{2k-1} \rangle$ and so $e_\beta(2k)$ acts on $F/\mathrm{Rad}(F)$ as $e_\beta(2k-2r)$. Hence u acts on $F/\mathrm{Rad}(F)$ as

$$(1 - e_\beta(2k-2r))(1 + e_\beta(2k-2r))^{-1} = V_\beta(2k-2r).$$

(ii) Let $r = 2$, so $F = V(1-u)^2 = \langle v_{-(2k-5)}, \ldots, v_{2k-1} \rangle$. Then $F^\perp = \langle v_{2k-3}, v_{2k-1} \rangle$, which is totally isotropic. □

Lemma 4.3.2 *Let $u = W(2l+1) \in \mathrm{Sp}(V)$ with $V = V_{4l+2}$, as defined in Sect. 4.1.1. Let $0 \leq r < 2l+1$, and $F = V(1-u)^r$.*

(i) *If $r \leq l$, then $u^{F/\mathrm{Rad}(F)} = W(2l-2r+1)$.*
(ii) *If $r > l$, then F is totally isotropic.*
(iii) *If $r = 2$ and $l \geq 2$, then $F^\perp$ is totally isotropic.*
(iv) *If $r = 2$ and $l = 1$, then $u^{F^\perp/\mathrm{Rad}(F^\perp)} = W(1)$.*
(v) *If $r = 2$ and $l = 0$, then $F = 0$.*

Proof As in the previous lemma, the proof is based on simple computations using the basis for $W(2l+1)$ and the nilpotent element e defined in Sect. 4.1.1. □

The Algorithm

Assume that $g \in G = \mathrm{Sp}(V)$ is unipotent and conjugate to

$$u = \bigoplus_{i=1}^{r} (V_{\beta_i}(2k_i) \oplus V_1(2k_i)^{[a_i-1]}) \oplus \bigoplus_{i=1}^{s} W(2l_i+1)^{[b_i]},$$

for some $\beta_i \in \{1, \alpha\}$. We now give an algorithm to compute the parameters β_i. To simplify notation, write

$$V_{\beta,a}(2k) = V_\beta(2k) \oplus V_1(2k)^{[a-1]}.$$

Subcase I. Assume that g is conjugate to $u = V_{\beta,a}(2) \oplus W(1)^{[b]}$. We compute β as follows. First, find $e \in V$ such that

$$(e,\ e(1-g)) \neq 0.$$

(The number of such vectors is $1 - O(\frac{1}{q})$.) Write $f = e(1-g)$ and $\gamma_1 = (e, f)$, and define $F = \langle e, f\rangle$. Then F is a non-degenerate g-invariant subspace, and with respect to the basis $[e, -\gamma_1^{-1} f]$ the matrix of the form is $\begin{pmatrix} & -1 \\ 1 & \end{pmatrix}$, and

$$g^F = \begin{pmatrix} 1 & -\gamma_1 \\ 0 & 1 \end{pmatrix}.$$

Hence g acts on F as $V_{\frac{1}{2}\gamma_1}(2)$. Now consider $F^{\perp}$ and repeat. Continuing, we compute values $\gamma_1, \ldots, \gamma_a$ such that g is conjugate to

$$W(1)^b \oplus \bigoplus_{i=1}^{a} V_{\frac{1}{2}\gamma_i}(2).$$

We conclude that $\beta \equiv 2^a \gamma_1 \cdots \gamma_a \bmod \mathbb{F}_q^2$.

General case. Now assume that g is conjugate to

$$u = \bigoplus_{i=1}^{r} V_{\beta_i,a_i}(2k_i) \oplus U,$$

where $k_1 < k_2 < \cdots < k_r$, and U is a direct sum of blocks of the form $W(2l_i + 1)$. We give an algorithm to compute the parameters β_i.

(1) Compute $F = V(1-g)^{k_1-1}$ and the action of g on $F_0 = F/\mathrm{Rad}(F)$. By Lemmas 4.3.1 and 4.3.2,

$$g^{F_0} = V_{\beta_1,a_1}(2) \oplus \bigoplus_{i=2}^{r} V_{\beta_i,a_i}(2k_i') \oplus U', \tag{4.3.2}$$

where U' is a direct sum of blocks of the form $W(2l_i + 1)$, and $k_i' = k_i - k_1 + 1$ (so $1 < k_2' < \cdots < k_r'$).

(2) Compute $F_1 = F_0(1-g)^2$ and the action of g on $F_2 = F_1^{\perp}/\mathrm{Rad}(F_1^{\perp})$. By the lemmas,

$$g^{F_2} = W(1)^{[b]} \oplus V_{\beta_1,a_1}(2).$$

Apply the algorithm of Subcase I to compute β_1.

(3) Consider g^{F_0} in (4.3.2). Compute $F_3 = F_0(1-g)$ and the action of g on $F_4 = F_3/\mathrm{Rad}(F_3)$. By the lemmas,

$$g^{F_4} = \bigoplus_{i=2}^{r} V_{\beta_i,a_i}(2k_i'') \oplus U'',$$

where U'' is a direct sum of blocks of the form $W(2l_i+1)$, and $k_i'' = k_i' - 1$. Now repeat Steps (1) and (2) to compute β_2.

Iterating, we compute all of the β_i.

4.3.2 Orthogonal Groups

Let $G = \mathrm{O}(V) \cong \mathrm{O}_n^{\epsilon}(q)$ with q odd, preserving an orthogonal form (,), and let $g \in G$ be unipotent. We want to compute the representative

$$\bigoplus_{i=1}^{r}(V_{\beta_i}(2k_i+1) \oplus V_1(2k_i+1)^{[a_i-1]}) \oplus \bigoplus_{i=1}^{s} W(2l_i)^{[b_i]}, \tag{4.3.3}$$

as defined in Sect. 4.1.2, such that g is G-conjugate to this representative. (Refining to conjugacy in $\mathrm{SO}(V)$ and $\Omega(V)$ is discussed at the end of this section.) The values $k_1, \ldots, k_r$ and $l_1, \ldots, l_s$ are determined by the Jordan form of g. It remains to determine the values of the parameters $\beta_i \in \{1, \alpha\}$ in (4.3.3).

Our algorithm to do this uses the following lemma.

Lemma 4.3.3

(a) *Let $u = V_\beta(2k+1) \in \mathrm{O}(V)$ with $V = V_{2k+1}$, as defined in Sect. 4.1.2. Let $0 \le r < k$, and $F = V(1-u)^r$.*
  (i) $u^{F/\mathrm{Rad}(F)} = V_\beta(2k-2r+1)$.
  (ii) *If $r = 1$, then $F^\perp$ is totally isotropic.*

(b) *Let $u = W(2l) \in \mathrm{O}(V)$ with $V = V_{4l}$, as defined in Sect. 4.1.2. Let $0 \le r < 2l$, and $F = V(1-u)^r$.*
  (i) *If $r \le l$, then $u^{F/\mathrm{Rad}(F)} = W(2l-2r)$.*
  (ii) *If $r > l$, then F is totally isotropic.*
  (iii) *If $r = 1$, then $F^\perp$ is totally isotropic.*

Proof As in the proof of Lemma 4.3.1, this is based on simple computations using the bases and nilpotent elements defining $V_\beta(2k+1)$ and $W(2l)$ in Sect. 4.1.2. □

The Algorithm

Assume that $g \in G = \mathrm{O}(V)$ is unipotent and conjugate to

$$u = \bigoplus_{i=1}^{r} (V_{\beta_i}(2k_i + 1) \oplus V_1(2k_i + 1)^{[a_i-1]}) \oplus U$$

where $k_1 < k_2 < \cdots < k_r$, and U is a direct sum of blocks of the form $W(2l_i)$. We now give an algorithm to compute the parameters β_i. To simplify notation, write $V_{\beta,a}(2k+1) = V_\beta(2k+1) \oplus V_1(2k+1)^{[a-1]}$.

(1) Compute $F = V(1-g)^{k_1}$ and the action of g on $F_0 = F/\mathrm{Rad}(F)$. By Lemma 4.3.3,

$$g^{F_0} = V_{\beta_1,a_1}(1) \oplus \bigoplus_{i=2}^{r} V_{\beta_i,a_i}(2k_i' + 1) \oplus U', \tag{4.3.4}$$

where $k_i' = k_i - k_1$.

(2) Compute $F_1 = F_0(1-g)$ and the action of g on $F_2 = F_1^\perp/\mathrm{Rad}(F_1^\perp)$. By Lemma 4.3.3,

$$g^{F_2} = V_{\beta_1,a_1}(1).$$

The determinant of the Gram matrix of this a_1-dimensional space is congruent to β_1 modulo $(\mathbb{F}_q)^2$. So this determinant determines β_1.

(3) Replace V by F_0, and g by g^{F_0} as in (4.3.4). Compute $F_3 = V(1-g)$ and the action of g on $F_3/\mathrm{Rad}(F_3)$, which is

$$g^{F_3/\mathrm{Rad}(F_3)} = \bigoplus_{i=2}^{r} V_{\beta_i,a_i}(2k_i'' + 1) \oplus U'',$$

where $k_i'' = k_i - 1$. Now repeat Steps (1) and (2) to compute β_2.

Iterating, we compute all of the β_i.

Conjugacy in SO(V) and $\Omega(V)$

As discussed in Sect. 4.1.2, a conjugacy class u^G in $G = \mathrm{O}(V)$ may split into two classes in $G_0 = \mathrm{SO}(V)$, with representatives u and u^t for a reflection t. For unipotent $g \in u^G$, we determine its G_0-class as follows: using the method of Sect. 4.4.2, compute $y \in G$ such that $g^y = u$. If $y \in G_0$, then g is G_0-conjugate to u; if $y \in G_0t$, then g is G_0-conjugate to u^t.

Similarly, an SO(V)-class u^{G_0} may split into two classes in $G_1 = \Omega(V)$, with representatives u and u^s where $s \in \mathrm{SO}(V) \setminus \Omega(V)$. For unipotent $g \in u^{G_0}$, we determine its G_1-class as follows: compute $y \in G_0$ such that $g^y = u$. Now determine which of G_1 or $G_1 s$ contains y.

4.3.3 *Unitary Groups*

Let $G = \mathrm{SU}_n(q)$. Let $u \in G$ be the matrix representative for $\bigoplus_{i=1}^s V(m_i)^{[r_i]}$ defined in Sect. 4.1.3. As described there, representatives of the G-conjugacy classes of elements with this Jordan form are $u^{d(\alpha)}$, where $d(\alpha) \in \mathrm{GU}_n(q)$ has determinant α, and α ranges over representatives of Z_{q+1}/Z_{q+1}^t where $t = \gcd(m_1, \ldots, m_s)$.

Our algorithm to construct the class representative in G of a unipotent element g is the following. Let g have Jordan form $\bigoplus_{i=1}^s J_{m_i}^{[r_i]}$, and let u be the matrix representative for $\bigoplus_{i=1}^s V(m_i)^{[r_i]}$. Using the work of Sect. 4.4.3, find $y \in \mathrm{GU}_n(q)$ such that $g^y = u$, and let $\det(g) = \beta \in Z_{q+1}$. Now g is G-conjugate to $u^{d(\alpha)}$ where α is congruent to β^{-1} modulo Z_{q+1}^t.

4.4 Constructing a Conjugating Element

In this section we complete our work for classical groups G in good characteristic by solving the following problem: given unipotent $g \in G$ that is conjugate to a class representative u, find $y \in G$ such that $g^y = u$.

4.4.1 *Symplectic Groups*

Let $G = \mathrm{Sp}(V) \cong \mathrm{Sp}_{2n}(q)$ with q odd. Recall from Sect. 4.1.1 that the unipotent class representatives in G are the elements

$$u = \bigoplus_{i=1}^{r} (V_{\beta_i}(2k_i) \oplus V_1(2k_i)^{[a_i-1]}) \oplus \bigoplus_{i=1}^{s} W(2l_i + 1)^{[b_i]}, \tag{4.4.1}$$

where $V = \bigoplus_{i=1}^r V_{2k_i}^{[a_i]} \oplus \bigoplus_{i=1}^s V_{4l_i+2}^{[b_i]}$, and each $\beta_i \in \{1, \alpha\}$ (and the k_i are distinct, as are the l_i). Given $g \in G$ that is conjugate to u, we aim to compute $y \in G$ such that $g^y = u$. We can compute y "block-by-block", so the main task is to solve the problem when u is a single block $V_\beta(2k)$ or $W(2l+1)$. We handle these cases separately.

Case 1: $u = V_\beta(2k)$

Let $u = V_\beta(2k) \in G = \mathrm{Sp}(V) \cong \mathrm{Sp}_{2k}(q)$. Recall from Sect. 4.1.1 that $u = (1 - e_\beta)(1 + e_\beta)^{-1}$, where e_β is a nilpotent element of $sp(V)$; moreover, if $v = v_{-(2k-1)}$, the first vector in the basis defining e_β, then V has a basis

$$v,\ ve_\beta,\ ve_\beta^2, \dots, ve_\beta^{k-1},\ \beta^{-1}ve_\beta^k, \dots, \beta^{-1}ve_\beta^{2k-1},$$

with respect to which the symplectic form defining G has matrix ϵB, where $\epsilon = (-1)^k$ and

$$B = \begin{pmatrix} & & & & 1 \\ & & & -1 & \\ & & 1 & & \\ & \iddots & & & \\ -1 & & & & \end{pmatrix}.$$

In terms of the basis,

$$(ve_\beta^i, ve_\beta^j) = \begin{cases} \epsilon\beta(-1)^i, & \text{if } i + j = 2k - 1 \\ 0, & \text{otherwise.} \end{cases}$$

Now suppose $g \in G$ is conjugate to u. We aim to compute $y \in G$ such that $g^y = u$. Let

$$f = (1 - g)(1 + g)^{-1},$$

a nilpotent element of $sp(V)$. We seek $w \in V$ with the property that

$$w,\ wf, \dots, wf^{k-1},\ \beta^{-1}wf^k, \dots, \beta^{-1}wf^{2k-1} \tag{4.4.2}$$

is a basis of V satisfying

$$(wf^i, wf^j) = \begin{cases} \epsilon\beta(-1)^i, & \text{if } i + j = 2k - 1 \\ 0, & \text{otherwise.} \end{cases} \tag{4.4.3}$$

Then the map sending $wf^i \mapsto ve_\beta^i$ for all i will lie in G and conjugate g to u.

Here is our algorithm to find such a vector w. First, we find $z \in V$ such that $(z, zf^{2k-1}) = \epsilon\beta$; this can be done by random selection. Now we aim to solve the following equation for $w \in V$ and $a_i \in \mathbb{F}_q$:

$$z = w \sum_{i=0}^{2k-1} a_i f^i \tag{4.4.4}$$

such that w satisfies (4.4.2) and (4.4.3).

Note that the fact that $(x, yf) + (xf, y) = 0$ for all $x, y \in V$ implies that for all i, j,

$$\begin{aligned}(zf^i, zf^j) &= (-1)^i (z, zf^{i+j}), \text{ and}\\ (z, zf^{2i}) &= 0.\end{aligned} \tag{4.4.5}$$

So, to ensure property (4.4.3) for w, we let $\alpha_j = (z, zf^{2j-1})$ for $1 \leq j \leq k$, and solve for a_i the equations

$$\begin{aligned}\alpha_j &= \left(\sum_r a_r wf^r, \sum_s a_s wf^{s+2j-1}\right)\\ &= \sum_{r+s+2j=2k} a_r a_s \epsilon\beta(-1)^r\end{aligned} \tag{4.4.6}$$

for $j = 1, \ldots, k$. These equations can be solved easily if we take $a_0 = 1$ and $a_i = 0$ for all odd i. So

$$z = w(1 + a_2 f^2 + a_4 f^4 + \cdots + a_{2k-2} f^{2k-2}), \tag{4.4.7}$$

and the equations (4.4.6) simplify to

$$\alpha_{k-i} = \epsilon\beta \sum_{r=0}^{i} a_{2i-2r} a_{2r} \quad (i = 0, \ldots, k-1).$$

For $i = 0$, the equation is $\alpha_k = \epsilon\beta a_0^2 = \epsilon\beta$, which is true by choice of z. For $i = 1$ the equation is $\alpha_{k-1} = 2\epsilon\beta a_2$, which we solve for a_2. Similarly we solve the $i = 2$ equation for a_4, and continuing, we determine a_{2i} for all i. Now let

$$m = 1 + a_2 f^2 + \cdots + a_{2k-2} f^{2k-2}.$$

Then m is invertible, and the vector $w = zm^{-1}$ has the required properties (4.4.2) and (4.4.3).

Summary of algorithm. Suppose $g \in G = \mathrm{Sp}_{2k}(q)$ is conjugate to $u = V_\beta(2k)$.

(1) Let $f = (1-g)(1+g)^{-1}$. Find $z \in V$ such that $(z, zf^{2k-1}) = \epsilon\beta$.
(2) Compute the values $\alpha_j = (z, zf^{2j-1})$ for $1 \leq j \leq k$. Solve (4.4.6) for the a_i, assuming that $a_0 = 1$ and $a_i = 0$ for all odd i.

(3) Let $m = 1 + a_2 f^2 + \cdots + a_{2k-2} f^{2k-2}$ and $w = zm^{-1}$. Define the linear map $y : wf^i \mapsto ve^i_\beta$ $(0 \le i \le 2k-1)$. Then $y \in G$ and $g^y = u$.

Case 2: $u = W(2l + 1)$

Let $u = W(2l+1) \in G = \mathrm{Sp}(V) \cong \mathrm{Sp}_{4l+2}(q)$. Recall from Sect. 4.1.1 that $u = (1-e)(1+e)^{-1}$, where e is a nilpotent element of $sp(V)$. If $w = w_{-2l}$ and $x = x_{-2l}$ are the first two vectors in the basis defining e, then V has a basis

$$w,\ we, \ldots, we^{2l}, x,\ xe, \ldots, xe^{2l},$$

where

$$(we^i,\ xe^j) = \begin{cases} \epsilon(-1)^i, & \text{if } i + j = 2l \\ 0, & \text{otherwise} \end{cases}$$

and $(we^i,\ we^j) = (xe^i,\ xe^j) = 0$ for all i, j. Also $\epsilon = (-1)^l$.

Now suppose $g \in G$ is conjugate to u. We aim to compute $y \in G$ such that $g^y = u$, and adopt a similar approach to the previous case. Let $f = (1-g)(1+g)^{-1}$, a nilpotent element of $sp(V)$. We seek $w', x' \in V$ with the property that

$$w',\ w'f, \ldots, w'f^{2l},\ x',\ x'f, \ldots, x'f^{2l} \tag{4.4.8}$$

is a basis of V satisfying

$$(w'f^i,\ x'f^j) = \begin{cases} \epsilon(-1)^i, & \text{if } i + j = 2l \\ 0, & \text{otherwise.} \end{cases} \tag{4.4.9}$$

and $(w'f^i,\ w'f^j) = (x'f^i,\ x'f^j) = 0$ for all i, j. Then the map sending $w'f^i \mapsto we^i, x'f^i \mapsto xe^i$ for all i will lie in G and conjugate g to u.

Here is our algorithm to find such vectors w', x'. First, we find $z, t \in V$ such that $\langle Vf,\ z,\ t\rangle = V$. Now we aim to solve the following equations for $w', x' \in V$ and $a_i, b_i, c_i, d_i \in \mathbb{F}_q$:

$$\begin{aligned} z &= w' \textstyle\sum_0^{2l} a_i f^i + x' \sum_0^{2l} b_i f^i, \\ t &= w' \textstyle\sum_0^{2l} c_i f^i + x' \sum_0^{2l} d_i f^i \end{aligned}$$

such that w', x' satisfy (4.4.8) and (4.4.9). In practice, we can find a solution in which many of the variables are assumed to be 0, namely a solution of the form

$$\begin{aligned} z &= w' \textstyle\sum_{i=0}^{2l} a_i f^i + x' \sum_{j=1}^{l} b_{2j-1} f^{2j-1}, \\ t &= w' \textstyle\sum_{i=1}^{l} c_{2i-1} f^{2i-1} + x'. \end{aligned} \tag{4.4.10}$$

To find this solution, we compute

$$\begin{aligned} \alpha_i &= (z,\ zf^{2i-1})\ (1 \le i \le l) \\ \beta_i &= (t,\ tf^{2i-1})\ (1 \le i \le l) \\ \gamma_j &= (z,\ tf^j) \quad\ (0 \le j \le 2l). \end{aligned} \tag{4.4.11}$$

Then for a_i, b_i, c_i as in (4.4.10), Eq. (4.4.9) implies:

$$\begin{aligned} \alpha_{l-i} &= 2\epsilon \textstyle\sum_{j=0}^{i} a_{2j} b_{2i-2j+1} && (i = 0, \dots, l-1) \\ \beta_{l-i} &= -2\epsilon c_{2i+1} && (i = 0, \dots, l-1) \\ \gamma_{2l-2i} &= \epsilon \left(a_{2i} + \textstyle\sum_{j=1}^{i} b_{2j-1} c_{2i-2j+1}\right) && (i = 0, \dots, l) \\ \gamma_{2l-2i+1} &= -\epsilon a_{2i-1} && (i = 1, \dots, l). \end{aligned} \tag{4.4.12}$$

These can be solved easily. First, we observe that

$$\begin{aligned} a_0 &= \epsilon \gamma_{2l}, \\ a_i &= -\epsilon \gamma_{2l-i} \text{ for } i \text{ odd}, \\ c_{2i-1} &= -\tfrac{1}{2}\epsilon \beta_{l-i+1} \text{ for } i = 1, \dots, l. \end{aligned}$$

Now we can solve successively for $b_1, a_2, b_3, a_4, \dots, b_{2l-1}, a_{2l}$. Note that $(z, tf^{2l}) \neq 0$ by choice of z, t, so $\gamma_{2l} \neq 0$ and hence $a_0 \neq 0$.

Given this solution for a_i, b_i, c_i, let

$$A = \sum_{i=0}^{2l} a_i f^i, \quad B = \sum_{i=1}^{l} b_{2i-1} f^{2i-1}, \quad C = \sum_{i=1}^{l} c_{2i-1} f^{2i-1}.$$

Then (4.4.10) is

$$\begin{aligned} z &= w'A + x'B, \\ t &= w'C + x' \end{aligned}$$

Solving these for w', x' (noting that $a_0 \neq 0$, so $A - CB$ is invertible), we obtain:

$$w' = (z - tB)(A - CB)^{-1}, \quad x' = t - w'C.$$

These vectors w', x' satisfy (4.4.8) and (4.4.9), as required.

Summary of Algorithm Suppose $g \in G = \mathrm{Sp}_{4l+2}(q)$ is conjugate to $u = W(2l+1)$.

(1) Let $f = (1-g)(1+g)^{-1}$. Find $z, t \in V$ such that $\langle Vf,\ z,\ t\rangle = V$.
(2) Compute the values $\alpha_i, \beta_i, \gamma_i$ as in (4.4.11). Solve (4.4.12) for the a_i, b_i, and c_i.

(3) Let $A = \sum_{i=0}^{2l} a_i f^i$, $B = \sum_{i=1}^{l} b_{2i-1} f^{2i-1}$, $C = \sum_{i=1}^{l} c_{2i-1} f^{2i-1}$, and let

$$w' = (z - tB)(A - CB)^{-1}, \quad x' = t - w'C.$$

Define the linear map

$$y : w' f^i \mapsto w e^i, \quad x' f^i \mapsto e^i \quad \text{for all } i.$$

Then $y \in G$ and $g^y = u$.

General Case

Suppose $g \in \mathrm{Sp}(V)$ is conjugate to a unipotent class representative

$$u = \bigoplus_{i=1}^{r} (V_{\beta_i}(2k_i) \oplus V_1(2k_i)^{[a_i-1]}) \oplus \bigoplus_{i=1}^{s} W(2l_i + 1)^{[b_i]}. \tag{4.4.13}$$

We find $y \in G$ such that $g^y = u$ as follows:

(a) Find a non-degenerate g-invariant subspace X on which the action of g is conjugate to $u_1 = V_\beta(m)$ or $W(m)$, where m is the largest Jordan block size in (4.4.13).
(b) By Case 1 or Case 2 above, compute $y_1 \in \mathrm{Sp}(X)$ that conjugates g^X to u_1.
(c) Now work in $X^\perp$ and repeat Steps (a) and (b).
(d) Continuing like this, obtain $y = \bigoplus y_i \in \mathrm{Sp}(V)$ conjugating g to u.

Note that in this process, when $2k_i$ becomes the largest block size, we must perform Step (a) a_i times, and we must ensure that the corresponding sequence of blocks $V_\beta(2k_i)$ obtained is $V_1(2k_i), \ldots, V_1(2k_i), V_{\beta_i}(2k_i)$.

It remains to give an algorithm to perform Step (a). Write $f = (1-g)(1+g)^{-1}$, a nilpotent element of $sp(V)$. Suppose first that the largest block size in (4.4.13) is $2l + 1$. By random choice, we can with high probability find $w, x \in V$ such that $(w, xf^{2l}) \neq 0$ and $(x, wf^{2l}) \neq 0$. Now set

$$X = \langle wf^i, xf^i : 0 \leq i \leq 2l \rangle.$$

Then X is g-invariant and non-degenerate, and g^X is conjugate to $W(2l + 1)$, as required. Now suppose the largest block size is $2k$. We find by random choice $v \in V$ such that $(v, vf^{2k-1}) \neq 0$. Now set

$$X = \langle vf^i : 0 \leq i \leq 2k - 1 \rangle.$$

Then X is non-degenerate and g^X is conjugate to $V_\beta(2k)$ for some $\beta \in \{1, \alpha\}$; we can compute β as in Sect. 4.3.1. As noted above, we require that the sequence of values of β for the blocks of size $2k_i + 1$ should be $1, \ldots, 1, \beta_i$; so, if necessary, then we re-choose the initial vector v in the above process until the correct value of β in this sequence is returned.

This completes our analysis of constructing conjugating elements in $Sp_{2n}(q)$ for odd q.

4.4.2 Orthogonal Groups

Let $G = \mathrm{O}(V) \cong \mathrm{O}_n^\epsilon(q)$ with q odd. Recall from Sect. 4.1.2 that the unipotent class representatives in G are the elements

$$u = \bigoplus_{i=1}^{r}(V_{\beta_i}(2k_i + 1) \oplus V_1(2k_i + 1)^{[a_i - 1]}) \oplus \bigoplus_{i=1}^{s} W(2l_i)^{[b_i]}, \tag{4.4.14}$$

where $V = \bigoplus_{i=1}^r V_{2k_i+1}^{[a_i]} \oplus \bigoplus_{i=1}^s V_{4l_i}^{[b_i]}$, and each $\beta_i \in \{1, \alpha\}$ (and the k_i are distinct, as are the l_i). Given $g \in G$ that is conjugate to u, we aim to compute $y \in G$ such that $g^y = u$. We then refine the method to handle conjugacy in $\mathrm{SO}(V)$ and $\Omega(V)$.

The approach used is similar to that of Sect. 4.4.1. As before, we can compute y "block-by-block", so the main task is to solve the problem when u is a single block $V_\beta(2k + 1)$ or $W(2l)$. We handle these cases separately.

Case 1: $u = V_\beta(2k + 1)$

Let $u = V_\beta(2k + 1) \in G = \mathrm{O}(V) \cong \mathrm{O}_{2k+1}(q)$, and let $e_\beta = (1 - u)(1 + u)^{-1}$, a nilpotent element of $so(V)$. If $v = v_{-2k}$, the first vector in the basis defining e_β, then V has a basis

$$v,\ ve_\beta,\ ve_\beta^2, \ldots, ve_\beta^k, \beta^{-1}ve_\beta^{k+1}, \ldots, \beta^{-1}ve_\beta^{2k},$$

with respect to which the orthogonal form defining G has values

$$(ve_\beta^i,\ ve_\beta^j) = \begin{cases} \epsilon\beta(-1)^i, & \text{if } i + j = 2k \\ 0, & \text{otherwise} \end{cases}$$

where $\epsilon = (-1)^k$.

Now suppose $g \in G$ is conjugate to u. We aim to compute $y \in G$ such that $g^y = u$. Let $f = (1-g)(1+g)^{-1}$, a nilpotent element of $so(V)$. We seek $w \in V$ with the property that $w, wf, \ldots, wf^k, \beta^{-1}wf^{k+1}, \ldots, \beta^{-1}wf^{2k}$ is a basis of V satisfying

$$(wf^i, wf^j) = \begin{cases} \epsilon\beta(-1)^i, & \text{if } i+j=2k \\ 0, & \text{otherwise.} \end{cases} \tag{4.4.15}$$

Then the map sending $wf^i \mapsto ve^i_\beta$ for all i will lie in G and conjugate g to u.

Here is our algorithm to find such a vector w. First, we find $z \in V$ such that $(z, zf^{2k}) = \epsilon\beta$. Now we aim to solve the following equation for $w \in V$ and $a_i \in \mathbb{F}_q$:

$$z = w\,(1 + a_2 f^2 + a_4 f^4 + \cdots + a_{2k} f^{2k}) \tag{4.4.16}$$

such that w satisfies (4.4.15).

Note that in the orthogonal case $(zf^i, zf^j) = (-1)^i(z, zf^{i+j})$, and $(z, zf^{2i+1}) = 0$ for all i, j. So to ensure property (4.4.15) for w, we let $\alpha_j = (z, zf^{2j})$ for $0 \leq j \leq k$, and solve for a_i the equations

$$\begin{aligned} \alpha_j &= \left(\textstyle\sum_r a_{2r} wf^{2r}, \sum_s a_{2s} wf^{2s+2j}\right) \\ &= \textstyle\sum_{r+s=k-j} a_{2r}a_{2s}\epsilon\beta \end{aligned} \tag{4.4.17}$$

for $j = 0, \ldots, k$. These equations can easily be solved uniquely for $a_2, a_4, \ldots, a_{2k}$. For this solution, let

$$m = 1 + a_2 f^2 + \cdots + a_{2k} f^{2k}.$$

Then m is invertible, and the vector $w = zm^{-1}$ has the required properties (4.4.15).

Case 2: $u = W(2l)$

Let $u = W(2l) \in G = \mathrm{O}(V) \cong \mathrm{O}^+_{4l}(q)$, and let $e = (1-u)(1+u)^{-1}$. If $w = w_{-(2l-1)}$ and $x = x_{-(2l-1)}$ are the first two vectors in the basis defining e, then V has a basis

$$w,\ we, \ldots, we^{2l-1}, x,\ xe, \ldots, xe^{2l-1},$$

where

$$(we^i, xe^j) = \begin{cases} \epsilon(-1)^i, & \text{if } i+j=2l-1 \\ 0, & \text{otherwise} \end{cases}$$

and $(we^i, we^j) = (xe^i, xe^j) = 0$ for all i, j.

Now suppose $g \in G$ is conjugate to u. We aim to compute $y \in G$ such that $g^y = u$. We let $f = (1-g)(1+g)^{-1}$, and seek $w', x' \in V$ with the property that $w', w'f, \ldots, w'f^{2l-1}, x', x'f, \ldots, x'f^{2l-1}$ is a basis of V satisfying

$$(w'f^i, x'f^j) = \begin{cases} \epsilon(-1)^i, & \text{if } i+j = 2l-1 \\ 0, & \text{otherwise} \end{cases} \tag{4.4.18}$$

and $(w'f^i, w'f^j) = (x'f^i, x'f^j) = 0$ for all i, j. Then the map sending $w'f^i \mapsto we^i, x'f^i \mapsto xe^i$ for all i will lie in G and conjugate g to u.

Here is our algorithm to find such vectors w', x'. First, we find $z, t \in V$ such that $\langle Vf, z, t\rangle = V$. In similar fashion to (4.4.10), we aim to solve the following equations for $w', x' \in V$ and $a_i, b_i, c_i \in \mathbb{F}_q$:

$$\begin{aligned} z &= w'\textstyle\sum_{i=0}^{2l-1} a_i f^i + x'\sum_{j=1}^{l} b_{2j-1} f^{2j-1}, \\ t &= w'\textstyle\sum_{i=1}^{l} c_{2i-1} f^{2i-1} + x' \end{aligned} \tag{4.4.19}$$

such that w', x' satisfy (4.4.18). To find this solution, we compute

$$\begin{aligned} \alpha_i &= (z, zf^{2i}) \ (0 \le i \le l-1) \\ \beta_i &= (t, tf^{2i}) \ \ (0 \le i \le l-1) \\ \gamma_j &= (z, tf^j) \ \ (0 \le j \le 2l-1). \end{aligned} \tag{4.4.20}$$

For a_i, b_i, c_i as in (4.4.19), Eq. (4.4.18) implies:

$$\begin{aligned} \alpha_{l-i} &= 2\epsilon \textstyle\sum_{j=0}^{i-1} a_{2j} b_{2i-2j+1} && (i = 1, \ldots, l) \\ \beta_{l-i} &= -2\epsilon c_{2i-1} && (i = 1, \ldots, l) \\ \gamma_{2l-2i-1} &= \epsilon\left(a_{2i} + \textstyle\sum_{j=1}^{i} b_{2j-1} c_{2i-2j+1}\right) && (i = 0, \ldots, l-1) \\ \gamma_{2l-2i} &= -\epsilon a_{2i-1} && (i = 1, \ldots, l). \end{aligned} \tag{4.4.21}$$

These equations can be solved easily for a_i, b_i, c_i. For this solution, let

$$A = \sum_{i=0}^{2l-1} a_i f^i, \quad B = \sum_{i=1}^{l} b_{2i-1} f^{2i-1}, \quad C = \sum_{i=1}^{l} c_{2i-1} f^{2i-1}.$$

Then the vectors

$$w' = (z - tB)(A - CB)^{-1}, \quad x' = t - w'C.$$

satisfy (4.4.18), as required.

General Case

Suppose $g \in \mathrm{O}(V)$ is conjugate to a unipotent class representative

$$u = \bigoplus_{i=1}^{r} (V_{\beta_i}(2k_i + 1) \oplus V_1(2k_i + 1)^{[a_i - 1]}) \oplus \sum_{i=1}^{s} W(2l_i)^{[b_i]}. \tag{4.4.22}$$

We find $y \in G$ such that $g^y = u$ as follows:

(a) Find a non-degenerate g-invariant subspace X on which the action of g is conjugate to $u_1 = V_\beta(m)$ or $W(m)$, where m is the largest Jordan block size in (4.4.22).
(b) By Case 1 or Case 2 above, compute $y_1 \in \mathrm{O}(X)$ that conjugates g^X to u_1.
(c) Now work in $X^\perp$ and repeat Steps (a) and (b).
(d) Continuing like this, obtain $y = \bigoplus y_i \in \mathrm{O}(V)$ conjugating g to u.

To perform Step (a), write $f = (1-g)(1+g)^{-1}$, a nilpotent element of $so(V)$. If the largest block size in (4.4.22) is $2l$, then find $w, x \in V$ such that $(w, xf^{2l-1}) \neq 0$ and $(x, wf^{2l-1}) \neq 0$, and set $X = \langle wf^i, xf^i : 0 \leq i \leq 2l - 1 \rangle$; now X is g-invariant and non-degenerate, and g^X is conjugate to $W(2l)$. If the largest block size is $2k + 1$, then find $v \in V$ such that $(v, vf^{2k}) \neq 0$ and set $X = \langle vf^i : 0 \leq i \leq 2k \rangle$; now X is g-invariant and non-degenerate, and g^X is conjugate to $V_\beta(2k + 1)$ for some $\beta \in \{1, \alpha\}$. We require that the sequence of values of β for the blocks of size $2k_i + 1$ should be $1, \ldots, 1, \beta_i$; so, if necessary, then we re-choose the initial vector v in the above process until the correct value of β in this sequence is returned.

This completes our analysis of constructing conjugating elements in $\mathrm{O}(V)$ (in odd characteristic).

Conjugation in $\mathrm{SO}(V)$ and $\Omega(V)$

We now address finding conjugating elements in $\mathrm{SO}(V)$ and $\Omega(V)$. Let u be as in (4.4.22).

First we consider $\mathrm{SO}(V)$. Assume that $u^{\mathrm{O}(V)}$ splits into two $\mathrm{SO}(V)$-classes, with representatives u, u^t as described in Sect. 4.1.2. Suppose g is $\mathrm{SO}(V)$-conjugate to one of these. To construct the conjugating element of $\mathrm{SO}(V)$, first compute $y \in \mathrm{O}(V)$ such that $g^y = u$. Now find $z \in \{1, t\}$ such that $yz \in \mathrm{SO}(V)$. Then $g^{yz} = u^z$, which is the required representative.

Now assume that $u^{\mathrm{O}(V)}$ is a single $\mathrm{SO}(V)$-class, so that $r \geq 1$ in (4.4.22). We can find $v \in C_{\mathrm{O}(V)}(u)$ of determinant -1: for example, $v = -I_{2k_i+1}$ acting on one of the $V(2k_i + 1)$ summands in (4.4.22). Suppose g is $\mathrm{SO}(V)$-conjugate to u, and compute $y \in \mathrm{O}(V)$ such that $g^y = u$. Then $g^y = g^{yv} = u$, and either y or yv is in $\mathrm{SO}(V)$.

Now consider $\Omega(V)$. Let $u \in \Omega(V)$, and assume first that $u^{\mathrm{SO}(V)}$ splits into two $\Omega(V)$-classes, with representatives u, u^s as described in Sect. 4.1.2. Let g be

$\mathrm{SO}(V)$-conjugate to u, and compute $y \in \mathrm{SO}(V)$ such that $g^y = u$. Now find $z \in \{1, s\}$ such that $yz \in \Omega(V)$. Then $g^{yz} = u^z$, which is the required representative.

Finally, suppose $u^{\mathrm{SO}(V)}$ is a single $\Omega(V)$-class. By [14, Prop. 2.4], $r \geq 1$ in (4.4.22) and there exists i such that one of the following holds:

(i) $a_i \geq 2$;
(ii) $\beta_i \not\equiv (-1)^{k_1+k_i}\beta_1 \bmod (\mathbb{F}_q)^2$.

In case (i), let X be a sum of two blocks of dimension $2k_i + 1$, and let u_X be the restriction of u to X. Now $C_{\mathrm{SO}(X)}(u_X)\,\Omega(X) = \mathrm{SO}(X)$. Compute representatives $\{1, s\}$ in $C_{\mathrm{SO}(X)}(u_X)$ for $\mathrm{SO}(X)/\Omega(X)$. Extend these to act trivially on $X^\perp$. Proceed as follows: suppose g is $\Omega(V)$-conjugate to u; compute $y \in \mathrm{SO}(V)$ such that $g^y = u$, and then find $z \in \{1, s\}$ such that $yz \in \Omega(V)$. Now $g^{yz} = u$, as required.

Case (ii) is similar. Let X be the subspace corresponding to the direct sum of the blocks $V_{\beta_1}(2k_1 + 1)$ and $V_{\beta_i}(2k_i + 1)$. The condition on β_i implies that $-1_X \notin \Omega(X)$. So if we let $s = -1_X$, then $\{1, s\}$ are representatives in $C_{\mathrm{SO}(X)}(u_X)$ for $\mathrm{SO}(X)/\Omega(X)$. Now proceed as in case (i).

4.4.3 Unitary Groups

Let $G = \mathrm{GU}(V) \cong \mathrm{GU}_n(q)$, and let

$$u = \bigoplus_{i=1}^{s} V(m_i)^{[r_i]}$$

be a unipotent class representative in G, as defined in Sect. 4.1.3. Given $g \in G$ that is conjugate to u, we aim to compute $y \in G$ such that $g^y = u$. Our approach is similar to that for the symplectic and orthogonal cases. Again we compute y "block-by-block". The analysis differs for blocks of even and odd size.

Case 1: $u = V(2k)$

Let $u = V(2k) \in G = \mathrm{GU}(V) \cong \mathrm{GU}_{2k}(q)$, and let $v = v_{-(2k-1)}$, the first vector in the basis defining u in Sect. 4.1.3. For the symplectic and orthogonal cases, we described the basis in terms of v and a single nilpotent element of the corresponding Lie algebra. Here, we use two nilpotent elements, namely e_1 and e_2, where

$$e_2 = u - 1, \quad e_1 = 1 - u^{-1} = e_2(1 + e_2)^{-1}.$$

Recall that $\beta \in \mathbb{F}_{q^2}^*$ satisfies $\beta + \bar{\beta} = 0$. The basis defining u as in Sect. 4.1.3 is

$$v,\ ve_1, \ldots, ve_1^{k-1}, \beta^{-1}ve_1^{k-1}e_2, -\beta^{-1}ve_1^{k-1}e_2^2, \ldots, (-1)^{k-1}\beta^{-1}ve_1^{k-1}e_2^k,$$

with respect to which the unitary form defining G has matrix

$$\begin{pmatrix} & & & 1 \\ & & 1 & \\ & \cdot^{\cdot^{\cdot}} & & \\ 1 & & & \end{pmatrix}. \tag{4.4.23}$$

Note that $e_1 = e_2(1+e_2)^{-1} = e_2(1-e_2+e_2^2-\cdots)$, so $e_1^k = e_1^{k-1}e_2(1-e_2+e_2^2-\cdots)$.

Now suppose $g \in G$ is conjugate to u. We aim to compute $y \in G$ such that $g^y = u$. Let $f_2 = g - 1$ and $f_1 = f_2(1 + f_2)^{-1}$. We seek a vector w such that

$$w,\ wf_1, \ldots, wf_1^{k-1}, \beta^{-1}wf_1^{k-1}f_2, -\beta^{-1}wf_1^{k-1}f_2^2, \ldots, (-1)^{k-1}\beta^{-1}wf_1^{k-1}f_2^k \tag{4.4.24}$$

is a basis of V with respect to which the unitary form has matrix (4.4.23). Then the map sending $wf_1^i \mapsto ve_1^i$, $wf_1^{k-1}f_2^i \mapsto ve_1^{k-1}e_2^i$ for all i will lie in G and conjugate g to u.

Our algorithm to find such a vector w is the following. First, we find $z \in V \setminus Vf_1$ such that $(z,\ \beta^{-1}zf_1^{k-1}f_2^k) = (-1)^{k+1}$. Now we aim to solve the following equation for $w \in V$ and $a_i, b_i \in \mathbb{F}_{q^2}$:

$$z = w\left(1 + \sum_{i=1}^{k-1} a_i f_1^i + \sum_{i=1}^{k} b_i \beta^{-1}(-1)^{i-1} f_1^{k-1} f_2^i\right), \tag{4.4.25}$$

such that w gives a basis (4.4.24) with respect to which the unitary form has matrix (4.4.23). Set

$$\begin{aligned} \alpha_i &= (z,\ zf_1^i) & (i = 0, \ldots, k-1) \\ \beta_i &= (z,\ \beta^{-1}(-1)^{i-1}zf_1^{k-1}f_2^i) & (i = 1, \ldots, k-1). \end{aligned}$$

To ensure that the matrix of the unitary form is (4.4.23), we need to solve $2k - 1$ equations for a_c, b_d; in each equation, the left hand side is α_i or β_i and the right hand side is a function of a_c, b_d. The simplest equation is

$$\beta_{k-1} = a_1 - \bar{a}_1.$$

We solve for a_1 using an additive Hilbert 90 algorithm. The next equation is

$$\beta_{k-2} = a_2 + \bar{a}_2 - \bar{a}_1 - a_1\bar{a}_1$$

This is solved for a_2. Now we proceed to the β_{k-3} equation, solving for a_3, and so on. In this way we can find a solution for a_c, b_d to all of the $2k-1$ equations. For this solution, define

$$m = 1 + \sum_{i=1}^{k-1} a_i f_1^i + \sum_{i=1}^{k} b_i \beta^{-1}(-1)^{i-1} f_1^{k-1} f_2^i.$$

Then $w = zm^{-1}$ is the required vector.

Case 2: $u = V(2k+1)$

Let $u = V(2k+1) \in G = \mathrm{GU}(V) \cong \mathrm{GU}_{2k+1}(q)$, and let $v = v_{-2k}$, the first vector in the basis defining u in Sect. 4.1.3. We use three nilpotent elements to describe the basis in terms of v: recall that $\gamma \in \mathbb{F}_{q^2}$ satisfies $\gamma + \bar{\gamma} = -1$, and define

$$e_2 = u - 1, \quad e_1 = 1 - u^{-1} = e_2(1+e_2)^{-1}, \quad e_0 = e_2(1 - \gamma e_2)^{-1}.$$

The basis defining u as in Sect. 4.1.3 is

$$v,\ ve_1, \ldots, ve_1^{k-1},\ ve_1^{k-1}e_0,\ -ve_1^{k-1}e_0e_2,\ ve_1^{k-1}e_0e_2^2, \ldots, (-1)^k ve_1^{k-1}e_0e_2^k,$$

with respect to which the unitary form defining G has matrix (4.4.23).

Now suppose $g \in G$ is conjugate to u. We aim to compute $y \in G$ such that $g^y = u$. Let $f_2 = g - 1$, $f_1 = f_2(1+f_2)^{-1}$ and $f_0 = f_2(1-\gamma f_2)^{-1}$. We seek a vector w such that

$$\begin{aligned} &w,\ wf_1, \ldots, wf_1^{k-1},\ wf_1^{k-1}f_0,\ -wf_1^{k-1}f_0f_2,\\ &wf_1^{k-1}f_0f_2^2, \ldots, (-1)^k wf_1^{k-1}f_0f_2^k, \end{aligned} \tag{4.4.26}$$

is a basis of V with respect to which the unitary form has matrix (4.4.23). Then the map sending $wf_1^i \mapsto ve_1^i$, $wf_1^{k-1}f_0f_2^i \mapsto ve_1^{k-1}e_0e_2^i$ for all i will lie in G and conjugate g to u.

Our algorithm to find such a vector w is similar to that of Case 1. First, we find $z \in V$ such that $(z, zf_1^{k-1}f_0f_2^k) = (-1)^k$. Now we aim to solve the following equation for $w \in V$ and $a_i, b_i \in \mathbb{F}_{q^2}$:

$$z = w\left(1 + \sum_{i=1}^{k-1} a_i f_1^i + \sum_{i=0}^{k} b_i (-1)^i f_1^{k-1} f_0 f_2^i\right), \tag{4.4.27}$$

such that w gives a basis (4.4.26) with respect to which the unitary form has matrix (4.4.23). To do this, set

$$\begin{aligned} \alpha_i &= (z, zf_1^i) \qquad\qquad (i = 0, \ldots, k-1) \\ \beta_i &= (z, zf_1^{k-1} f_0 f_2^i) \; (i = 0, \ldots, k-1). \end{aligned}$$

This leads to $2k$ equations for a_i, b_i, the simplest of which is the β_{k-1}-equation, which is $\beta_{k-1} = (-1)^k(\bar{a}_1 - a_1)$. We solve for a_1, then proceed to the β_{k-2}-equation, and so on. Having solved for all a_i, b_i, we define

$$m = 1 + \sum_{i=1}^{k-1} a_i f_1^i + \sum_{i=0}^{k} b_i (-1)^i f_1^{k-1} f_0 f_2^i.$$

Then $w = zm^{-1}$ is the required vector.

General Case

Suppose $g \in \mathrm{GU}(V)$ is conjugate to a unipotent class representative

$$u = \bigoplus_{i=1}^{s} V(m_i)^{[r_i]}, \tag{4.4.28}$$

where $m_1 > m_2 > \cdots > m_s$. As before, we find $y \in G$ such that $g^y = u$ proceeding "block-by-block". The main step is to compute a largest non-degenerate block for g, starting with $V(m_1)$. To do this, we find $v \in V$ such that $(v, v(g-1)^{m_1 - 1}) \neq 0$ and set

$$X = \langle v(g-1)^i : 0 \leq i \leq m_1 - 1 \rangle.$$

Then X is non-degenerate and g^X is conjugate to $V(m_1)$. Now, as in Case 1 or 2 above, we compute $y_1 \in \mathrm{GU}(X)$ conjugating g^X to the chosen representative for $V(m_1)$, and repeat the process in $X^{\perp}$, ending up with $y = \bigoplus y_i \in \mathrm{GU}(V)$ that conjugates g to u.

Conjugation in SU(V)

We now address the issue of finding conjugating elements in $\mathrm{SU}(V)$. Let $G = \mathrm{SU}(V) \cong \mathrm{SU}_n(q)$. Recall from Sect. 4.1.3 that our representatives for the unipotent classes of G are $u^{d(\alpha)}$, where u is as in (4.4.28), $d(\alpha) \in \mathrm{GU}(V)$ has determinant α, and α ranges over representatives of Z/Z^t, where $Z = \{\mu \in \mathbb{F}_{q^2} : \mu\bar{\mu} = 1\}$ and $t = \gcd(m_1, \ldots, m_s)$.

Suppose $g \in G$ is G-conjugate to $u^{d(\alpha)}$. As above, we can find $y \in \mathrm{GU}(V)$ such that $g^y = u$. Let $\beta = \det(y)$. If $\beta^{-1}\alpha^{-1} \in Z^t$, then we can find $c \in C_{\mathrm{GU}(V)}(u)$ of determinant $\beta^{-1}\alpha^{-1}$ (acting as a scalar on each block). Now $g^{ycd(\alpha)} = u^{d(\alpha)}$, and the conjugating element $ycd(\alpha)$ has determinant 1; hence it lies in G as required.

4.5 Some Examples

We close this chapter by listing in Tables 4.1, 4.2 and 4.3 the unipotent class representatives and centralizer orders for the 7-dimensional orthogonal groups and the 8-dimensional orthogonal and symplectic groups over fields of odd size q. The structures of the centralizers are given by Theorems 4.1.2 and 4.1.3.

Notation is as in earlier sections. Recall that $\alpha \in \mathbb{F}_q$ is a fixed non-square, and both β and β_i are in $\{1, \alpha\}$. The discriminant of the orthogonal form fixed by $G = \mathrm{SO}_8^\epsilon(q)$ is a square in $\mathbb{F}_q$ if and only if $\epsilon = +$ (see [21, Prop. 2.5.13]). In Table 4.3 we use the notation Δ for the value of the discriminant modulo $(\mathbb{F}_q^*)^2$; column 3 lists the conditions which determine the sign $\epsilon = \pm$ of the group $\mathrm{SO}_8^\epsilon(q)$ containing u. Tables 4.1 and 4.3 have a column indicating which unipotent classes split in the corresponding group $\Omega_7(q)$ or $\Omega_8^\epsilon(q)$; this information is recorded in Sect. 4.1.2.

Table 4.1 Unipotent class representatives in $G = \mathrm{SO}_7(q)$, q odd

Representative u	$\lvert C_G(u)\rvert$	Class splits in $\Omega_7(q)$?
$V_1(1)^{[7]}$	$\lvert G\rvert$	No
$W(2) \oplus V_1(1)^{[3]}$	$q^7 \lvert\mathrm{Sp}_2(q)\rvert\, \lvert\mathrm{SO}_3(q)\rvert$	No
$W(2) \oplus V_1(3)$	$q^6 \lvert\mathrm{Sp}_2(q)\rvert$	Yes
$V_\beta(3) \oplus V_\beta(1) \oplus V_1(1)^{[3]}$ $(\beta \in \{1, \alpha\})$	$q^5 \lvert\mathrm{O}_4^\pm(q)\rvert$	No
$V_\beta(3) \oplus V_1(3) \oplus V_\beta(1)$ $(\beta \in \{1, \alpha\})$	$2q^6(q \pm 1)$	No
$V_\beta(5) \oplus V_\beta(1) \oplus V_1(1)$ $(\beta \in \{1, \alpha\})$	$2q^4(q \pm 1)$	No
$V_1(7)$	q^3	Yes

Table 4.2 Unipotent class representatives in $G = \mathrm{Sp}_8(q)$, q odd

Representative u	$\lvert C_G(u)\rvert$
$W(1)^{[4]}$	$\lvert G\rvert$
$W(3) \oplus W(1)$	$q^8\lvert\mathrm{Sp}_2(q)\rvert^2$
$W(3) \oplus V_\beta(2)$ $(\beta \in \{1, \alpha\})$	$2q^9\lvert\mathrm{Sp}_2(q)\rvert$
$W(1)^{[3]} \oplus V_\beta(2)$ $(\beta \in \{1, \alpha\})$	$2q^7\lvert\mathrm{Sp}_6(q)\rvert$
$W(1)^{[2]} \oplus V_\beta(2) \oplus V_1(2)$ $(\beta \in \{1, \alpha\})$	$2q^{11}\lvert\mathrm{Sp}_4(q)\rvert(q \pm 1)$
$W(1)^{[2]} \oplus V_\beta(4)$ $(\beta \in \{1, \alpha\})$	$2q^6\lvert\mathrm{Sp}_4(q)\rvert$
$W(1) \oplus V_\beta(2) \oplus V_1(2)^{[2]}$ $(\beta \in \{1, \alpha\})$	$q^{12}\lvert\mathrm{Sp}_2(q)\rvert\,\lvert\mathrm{O}_3(q)\rvert$
$W(1) \oplus V_\beta(4) \oplus V_\gamma(2)$ $(\beta, \gamma \in \{1, \alpha\})$	$4q^9\lvert\mathrm{Sp}_2(q)\rvert$
$W(1) \oplus V_\beta(6)$ $(\beta \in \{1, \alpha\})$	$2q^5\lvert\mathrm{Sp}_2(q)\rvert$
$V_\beta(2) \oplus V_1(2)^{[3]}$ $(\beta \in \{1, \alpha\})$	$q^{10}\lvert\mathrm{O}_4^\pm(q)\rvert$
$V_\beta(4) \oplus V_\gamma(2) \oplus V_1(2)$ $(\beta, \gamma \in \{1, \alpha\})$	$4q^9(q \pm 1)$ (2 of each size)
$V_\beta(4) \oplus V_1(4)$ $(\beta \in \{1, \alpha\})$	$2q^7(q \pm 1)$
$V_\beta(6) \oplus V_\gamma(2)$ $(\beta, \gamma \in \{1, \alpha\})$	$4q^6$
$V_\beta(8)$ $(\beta \in \{1, \alpha\})$	$2q^4$

Table 4.3 Unipotent class representatives in $G = \mathrm{SO}_8^\epsilon(q)$, q odd

Representative u	$\lvert C_G(u)\rvert$	Condition	Class splits in $\Omega_8^\epsilon(q)$?
$V_\beta(1) \oplus V_1(1)^{[7]}$	$\lvert G\rvert$	$\beta \equiv \Delta \bmod (\mathbb{F}_q^*)^2$	No
$W(4)$, $W(4)^t$ (t reflection)	$q^5\lvert\mathrm{Sp}_2(q)\rvert$	$\epsilon = +$	Yes
$W(2)^{[2]}$, $W(2)^t \oplus W(2)$	$q^6\lvert\mathrm{Sp}_4(q)\rvert$	$\epsilon = +$	Yes
$W(2) \oplus V_\beta(1) \oplus V_1(1)^{[3]}$	$q^9\lvert\mathrm{Sp}_2(q)\rvert\,\lvert\mathrm{SO}_4^\epsilon(q)\rvert$	$\beta \equiv \Delta$	No
$W(2) \oplus V_{\beta_1}(3) \oplus V_{\beta_2}(1)$	$2q^9\lvert\mathrm{Sp}_2(q)\rvert$	$-\beta_1\beta_2 \equiv \Delta$	Yes, $\epsilon = +$ No, $\epsilon = -$
$V_{\beta_1}(3) \oplus V_{\beta_2}(1) \oplus V_1(1)^{[4]}$	$2q^6\lvert\mathrm{SO}_5(q)\rvert$	$-\beta_1\beta_2 \equiv \Delta$	No
$V_{\beta_1}(3) \oplus V_1(3) \oplus V_{\beta_2}(1) \oplus V_1(1)$	$2q^8(q \pm 1)^2$, $\epsilon = +$ $2q^8(q^2 - 1)$, $\epsilon = -$	$-\beta_1\beta_2 \equiv \Delta$	No
$V_{\beta_1}(5) \oplus V_{\beta_2}(1) \oplus V_1(1)^{[2]}$	$2q^5\lvert\mathrm{SO}_3(q)\rvert$	$\beta_1\beta_2 \equiv \Delta$	No
$V_{\beta_1}(5) \oplus V_{\beta_2}(3)$	$2q^6$	$-\beta_1\beta_2 \equiv \Delta$	Yes, $\epsilon = +$ No, $\epsilon = -$
$V_{\beta_1}(7) \oplus V_{\beta_2}(1)$	$2q^4$	$-\beta_1\beta_2 \equiv \Delta$	Yes, $\epsilon = +$ No, $\epsilon = -$

Chapter 5
Unipotent Classes in Bad Characteristic

In this chapter we solve the conjugacy problems (1)–(3) of Sect. 1.1 for unipotent elements of symplectic and orthogonal groups in characteristic 2. Many of the methods of Chap. 4 do not apply: there is no Cayley map between unipotent and nilpotent elements, nor is there a factorization of the centralizer of a unipotent element in a canonical parabolic subgroup. Hence, although we can exploit some features of the good characteristic case, our algorithms are more complicated.

5.1 Unipotent Class Representatives

We begin by describing the conjugacy class representatives. Let $G = \mathrm{Sp}(V)$ or $\mathrm{O}(V)$, where $V = V_{2n}(q)$, a vector space of dimension $2n$ over $\mathbb{F}_q$ and $q = 2^a$. Denote by $(\ ,\ )$ a symplectic form on V preserved by G, and by Q a quadratic form in the orthogonal case.

The representatives are defined as orthogonal direct sums of various indecomposable blocks, as defined in [14] (with a correction for the block $W_\beta(2l+1)$, see Sect. 5.1.3). These blocks are denoted $V_\beta(2k)$, $W(k)$ and $W_\beta(2l+1)$ for $\beta \in \{0, \alpha\}$, where α is a fixed element of $\mathbb{F}_q$ such that the quadratic $x^2+x+\alpha$ is irreducible over $\mathbb{F}_q$. We define these blocks, using bases labelled by vectors $v_i^{(j)}$, $w_i^{(j)}$, $x_i^{(j)}$; these will be used to define a certain parabolic subgroup that contains the centralizer of the unipotent element, as in Sect. 4.1.

G. De Franceschi et al., *Conjugacy in Finite Classical Groups*, Springer Monographs in Mathematics, https://doi.org/10.1007/978-3-031-86461-2_5

5.1.1 Blocks $V_\beta(2k)$

Let $k \geq 1$ and let V_{2k} be a $2k$-dimensional vector space over $\mathbb{F}_q$, with basis

$$v_{-(2k-1)}, v_{-(2k-3)}, \ldots, v_{2k-1}.$$

Define a symplectic form on V_{2k} by setting $(v_i, v_{-i}) = 1$ for all i, and all other values $(v_i, v_j) = 0$. For $k \geq 2$, define $V_\beta(2k)$ to be the linear map acting as follows:

$$\begin{aligned} v_i &\mapsto v_i + v_{i+2} + \cdots + v_1 + \beta v_3 \text{ for } -(2k-1) \leq i \leq -3, \\ v_i &\mapsto v_i + v_{i+2} \qquad\qquad\qquad\;\; \text{for } -1 \leq i < 2k-1, \\ v_{2k-1} &\mapsto v_{2k-1}. \end{aligned}$$

For $k = 1$, define $V_\beta(2)$ as follows: $v_{-1} \mapsto v_{-1} + v_1$ and $v_1 \mapsto v_1$.

Observe that $V_\beta(2k)$ lies in $\mathrm{Sp}(V_{2k})$ and acts as a single Jordan block. If we define a quadratic form Q_β on V associated to (,) by setting $Q_\beta(v_{-1}) = \beta$, $Q_\beta(v_1) = 1$ and Q-values 0 on all other basis vectors, then $V_\beta(2k)$ is in the orthogonal group preserving Q_β. Specifically, $V_0(2k) \in \mathrm{O}^+_{2k}(q)$ and $V_\alpha(2k) \in \mathrm{O}^-_{2k}(q)$. For $\beta = 0$ we write just $V(2k)$ instead of $V_0(2k)$. For all these assertions, see [14, §3].

5.1.2 Blocks $W(k)$

Let $k \geq 1$ and let V_{2k} be a $2k$-dimensional vector space over $\mathbb{F}_q$, with basis

$$w_{-(k-1)}, x_{-(k-1)}, w_{-(k-3)}, x_{-(k-3)}, \ldots, w_{k-1}, x_{k-1}.$$

Define a symplectic form on V_{2k} by setting $(w_i, x_{-i}) = 1$ for all i, and all other values (w_i, w_j), (w_i, x_j), (x_i, x_j) equal 0; and let Q be the quadratic form associated to (,) such that $Q(w_i) = Q(x_i) = 0$ for all i. Now define $W(k)$ to be the linear map acting as follows:

$$\begin{aligned} w_i &\mapsto w_i + w_{i+2} \qquad\qquad\quad\;\; \text{for } -(k-1) \leq i < k-1, \\ w_{2k-1} &\mapsto w_{2k-1}, \\ x_i &\mapsto x_i + x_{i+2} + \cdots + x_{k-1} \text{ for } -(k-1) \leq i \leq k-1. \end{aligned}$$

Then $W(k)$ lies in $\mathrm{Sp}(V_{2k})$, and also in $\Omega^+(V_{2k})$, the orthogonal group of plus type preserving Q. It acts as a sum of two Jordan blocks of size k.

5.1.3 Blocks $W_\beta(2l+1)$

Let $l \geq 1$ and let V_{4l+2} be a $(4l+2)$-dimensional vector space over $\mathbb{F}_q$, with basis

$$w_{-2l}, x_{-2l}, w_{-(2l-2)}, x_{-(2l-2)}, \ldots, w_{2l}, x_{2l}.$$

Define a symplectic form on V_{4l+2} as in Sect. 5.1.2, and let Q'_β be the quadratic form associated to $(\ ,\ )$ such that $Q'_\beta(w_{-2}) = Q'_\beta(w_0) = Q'_\beta(x_0) = \beta$, and Q'_β takes the value 0 on all other basis vectors. Now define $W_\beta(2l+1)$ to be the linear map acting as follows:

$$\begin{aligned}
w_i &\mapsto w_i + w_{i+2} && \text{for } i \neq -4, -2, 2l,\\
w_{-4} &\mapsto w_{-4} + w_{-2} + \beta x_2, &&\\
w_{-2} &\mapsto w_{-2} + w_0 + \beta x_2, &&\\
w_{2l} &\mapsto w_{2l}, &&\\
x_i &\mapsto x_i + x_{i+2} + \cdots + x_{2l} + \beta w_2 && \text{for } -2l \leq i \leq -2,\\
x_i &\mapsto x_i + x_{i+2} + \cdots + x_{2l} && \text{for } 0 \leq i \leq 2l-2,\\
x_{2l} &\mapsto x_{2l}.
\end{aligned}$$

This is a correction to the block $W_\beta(2l+1)$ as defined in [14]. Note that $W_0(2l+1)$ is the element $W(2l+1)$ of $\Omega^+_{4l+2}(q)$ defined in Sect. 5.1.2, and $W_\alpha(2l+1) \in \Omega^-_{4l+2}(q)$.

5.1.4 The General Case

By [14, Thm. 3.1], every unipotent element of $G = \mathrm{Sp}(V)$ or $\mathrm{O}(V)$ is conjugate to exactly one element of the form

$$\bigoplus_i W(m_i)^{[c_i]} \oplus \bigoplus_j V(2k_j)^{[d_j]} \oplus \bigoplus_r W_\alpha(m'_r) \oplus \bigoplus_s V_\alpha(2k'_s) \qquad (5.1.1)$$

satisfying the following conditions:

(i) the m'_r are odd and distinct, and the k'_s are distinct;
(ii) $d_j \leq 2$, and $d_j \leq 1$ if there exist j, s such that $k_j = k'_s$;
(iii) there exist no j, s such that $k'_s - k_j = 1$ or $k'_s - k'_j = 1$;
(iv) there exist no j, r such that $m'_r = 2k_j \pm 1$ or $m'_r = 2k'_j \pm 1$;
(v) for $G = \mathrm{Sp}(V)$, each $m'_r \geq 3$ and each $k'_s \geq 2$.

In the orthogonal case, the element is in $\Omega(V)$ if and only if the total number of V- and V_α-blocks is even; moreover, the only G-classes which split into two $\Omega(V)$-classes are those of the form $\bigoplus W(m_i)^{[a_i]}$ with all m_i even, and for these a second class representative can be obtained by replacing one summand $W(m_i)$ by $W(m_i)^t$ with t a reflection.

5.1.5 Class Representatives

We choose a matrix representative for the element (5.1.1) in a manner similar to Sect. 4.1. List the V-blocks and W-blocks in (5.1.1) in decreasing order of Jordan block sizes. For the ith space, if it is $V_\beta(2k)$, then label its basis as

$$v^{(i)}_{-(2k-1)}, v^{(i)}_{-(2k-3)}, \ldots, v^{(i)}_{2k-1};$$

if it is $W(k)$ or $W_\beta(k)$, then label its basis as

$$w^{(i)}_{-(k-1)}, x^{(i)}_{-(k-1)}, \ldots, w^{(i)}_{k-1}, x^{(i)}_{k-1}.$$

Take a basis of V consisting of these vectors $v_j^{(i)}, w_j^{(i)}, x_j^{(i)}$ in increasing order lexicographically with respect to the pairs (j, i). Choose the matrix representative corresponding to the linear map (5.1.1) to be its matrix with respect to this ordered basis.

5.2 Centralizers of Class Representatives

We now discuss how to construct the centralizers of these unipotent class representatives. We begin by describing the structure of the centralizer of a representative (5.1.1); this is given by [22, Thm. 7.3].

To make the description easier, we rewrite (5.1.1) slightly differently. For m_i odd, let a_i be the total number of W-blocks of size m_i (including $W_\alpha(m_i)$ if it is present), and write

$$W_{\beta,a_i}(m_i) = \begin{cases} W(m_i)^{[a_i]}, & \text{if } \beta = 0 \\ W(m_i)^{[a_i-1]} \oplus W_\alpha(m_i), & \text{if } \beta = \alpha. \end{cases}$$

Similarly, for the V-blocks define

$$V_{\beta,b_i}(2k_i) = \begin{cases} V(2k_i)^{[b_i]}, & \text{if } \beta = 0 \\ V(2k_i)^{[b_i-1]} \oplus V_\alpha(2k_i), & \text{if } \beta = \alpha. \end{cases}$$

Now rewrite (5.1.1) as

$$\bigoplus_{m_i \text{ even}} W(m_i)^{[a_i]} \oplus \bigoplus_{m_i \text{ odd}} W_{\beta_i, a_i}(m_i) \oplus \bigoplus_j V_{\beta_j, b_j}(2k_j). \tag{5.2.1}$$

Thus $a_i = c_i$ or $c_i + 1$, where c_i is as in (5.1.1); and $a_i = c_i + 1$ if and only if there is a summand $W_\alpha(m_i)$. Similarly, $b_j = d_j$ or $d_j + 1$.

Let u be the unipotent element (5.2.1), and let $\mathcal{J}_u$ be the set of Jordan block sizes in the Jordan form of u. Define a function $\chi_u : \mathcal{J}_u \to \mathbb{N}$ as follows:

$$G = \mathrm{Sp}(V) : \chi_u(n) = \begin{cases} \frac{1}{2}n, & \text{if } \exists\, V(n) \text{ or } V_\alpha(n) \text{ in } (5.1.1) \\ \lfloor \frac{1}{2}(n-1) \rfloor, & \text{otherwise.} \end{cases}$$

$$G = \mathrm{O}(V) : \ \chi_u(n) = \begin{cases} \frac{1}{2}(n+2), & \text{if } \exists\, V(n) \text{ or } V_\alpha(n) \text{ in } (5.1.1) \\ \lfloor \frac{1}{2}(n+1) \rfloor, & \text{otherwise.} \end{cases}$$

Let $f_1, \ldots, f_s$ be the list of Jordan block sizes in (5.2.1), including multiplicities, where $f_1 \geq f_2 \geq \cdots \geq f_s$, and define

$$N_u = \sum_{i=1}^{s} (i f_i - \chi_u(f_i)). \tag{5.2.2}$$

Then N_u is the dimension of the centralizer of u in the algebraic group $\bar{G} = \mathrm{Sp}(\bar{V})$ or $\mathrm{SO}(\bar{V})$ over $\bar{\mathbb{F}}_q$ corresponding to G, where $\bar{V} = V \otimes \bar{\mathbb{F}}_q$ (see [22, (4.1)]).

The following is [22, Thm. 7.3].

Theorem 5.2.1 *Let $G = \mathrm{Sp}(V)$ or $\mathrm{O}(V)$, and let u be as in (5.2.1); in the orthogonal case, assume $u \in \Omega(V)$. Let $C = C_G(u)$.*

(i) *C has a normal subgroup R of order q^{M_u}, where M_u is given in part (iii).*
(ii)

$$C/R \cong \prod_{m_i \text{ even}} \mathrm{Sp}_{2a_i}(q) \times \prod_{m_i \text{ odd}} I_{2a_i}(q) \times Z_2^{t+\delta}, \tag{5.2.3}$$

where

$$I_{2a_i}(q) = \begin{cases} \mathrm{Sp}_{2a_i}(q), & \text{if } m_i \pm 1 = 2k_j \text{ or } 2k'_j \text{ for some } j, \\ \mathrm{Sp}_{2a_i}(q), & \text{if } m_i = 1 \text{ and } G = \mathrm{Sp}(V), \\ \mathrm{O}_{2a_i}^{\epsilon_i}(q), & \text{otherwise;} \end{cases}$$

and t is the number of values of j such that $k_j - k_{j+1} \geq 2$ where we list the k_j in (5.2.1) in decreasing order; and $\delta \in \{0, 1\}$ with $\delta = 0$ if and only if either there are no V-blocks, or $G = \mathrm{Sp}(V)$ and there is a block $V(2)$.

(iii) $|R| = q^{M_u}$, *where*

$$M_u = N_u - \sum_{m_i \text{ even}} \dim \mathrm{Sp}_{2a_i} - \sum_{m_i \text{ odd}} \dim I_{2a_i}.$$

For the remainder of this section G denotes either $\mathrm{Sp}(V)$ or $\Omega(V)$ (rather than $\mathrm{O}(V)$). Let F be a Frobenius endomorphism of $\bar{G}$ such that $G = \bar{G}^F$ (see [22, §2.4]). Note that the normal subgroup R in part (i) is the fixed point group under F of the unipotent radical of $C_{\bar{G}}(u)$.

As in good characteristic, the key to computing $C_G(u)$ is to construct a canonical parabolic subgroup that contains this centralizer. We define this parabolic subgroup to be the stabilizer of a flag defined by the "weights" of the basis vectors, as we now describe. Note that we do not define a 1-dimensional torus (as we did in the previous chapter), so these weights are just a notational device, and are not the weights of a torus.

First, we tweak the bases for the blocks $W_\alpha(2l+1)$ for the orthogonal group. For such a block, with basis $w_{-2l}, x_{-2l}, \ldots, w_{2l}, x_{2l}$, we change the basis vectors w_{-2} and x_0 as follows:

$$w_{-2} \mapsto w_{-2} + w_0, \quad x_0 \mapsto x_0 + x_2.$$

With this change, the values of the quadratic form Q'_β are nonzero only for the basis vectors w_0 and x_0.

Now we define a canonical parabolic subgroup associated to u, the representative (5.2.1) in $G = \mathrm{Sp}(V)$ or $\Omega(V)$. For a basis vector $v_j^{(i)}$, $w_j^{(i)}$ or $x_j^{(i)}$, the subscript j is its *weight*; the span of all basis vectors of weight j is the *weight space* for j. Define P to be the parabolic subgroup stabilizing the flag of V defined by sums of weight spaces for decreasing weights. Write $P = QL$, with unipotent radical Q and Levi subgroup L.

The next result follows from [8, Thm. 5.3].

Theorem 5.2.2 *The centralizer in G of u is a subgroup of P.*

Unlike the good characteristic case, it is not true in general that $C_P(u) = C_Q(u)C_L(u)$. From Theorem 5.2.1(i) we know that $C_G(u)$ has a normal subgroup R of order q^{M_u}. We conjecture the order of $RC_Q(u)/C_Q(u)$.

Conjecture 5.2.3 Let u be as in (5.2.1), and let $J = \{j : b_j = 2\}$. For $j \in J$, let a_j be the multiplicity of $W(2k_j)$ in (5.2.1). Then $|RC_Q(u)/C_Q(u)| = q^t$, where $t = \sum_{j \in J}(2a_j + 1)$.

In our algorithm given below to construct the centralizer, we assume the conjecture. In any given example, it can be verified by checking that the subgroup R of $C_G(u)$ constructed in (e) below has order q^{M_u}, as given in Theorem 5.2.1(i).

We construct $C_G(u)$ in the following steps:

(a) construct $C_Q(u)$;
(b) construct subgroups of $C_G(u)$ that cover the factors $\mathrm{Sp}_{2a_i}(q)$ in (5.2.3);
(c) construct subgroups of $C_G(u)$ that cover the factors $I_{2a_i}(q)$ in (5.2.3);
(d) construct the subgroup of $C_G(u)$ that covers the factor $Z_2^{t+\delta}$ in (5.2.3);
(e) construct the subgroup R of $C_G(u)$.

Each step typically requires more effort than the corresponding step in good characteristic.

5.2.1 *Constructing* $C_Q(u)$

As in good characteristic, the main task is to write down a generating set for the unipotent group Q. As discussed in Sect. 4.2, we then construct a power-conjugate presentation for Q and so construct $C_Q(u)$.

Generators for Q

For $G = \mathrm{Sp}(V)$, we can use exactly the generators given in Sect. 4.2.1 (the characteristic is 2, so the sign ϵ can be ignored).

Now consider $G = \Omega(V)$. The generators for Q differ from those in good characteristic. We define ten collections of generators for Q.

(1) For all $i > 0$, all k, l, and every $\lambda \in \mathbb{F}_q$ define the following generators which fix all basis vectors other than those listed.

$$\begin{aligned} x_{1ikl}(\lambda) : w_{-i}^{(k)} &\mapsto w_{-i}^{(k)} + \lambda x_0^{(l)} + \lambda^2 Q(x_0^{(l)}) x_i^{(k)}, \\ w_0^{(l)} &\mapsto w_0^{(l)} + \lambda x_i^{(k)} \end{aligned}$$

$$\begin{aligned} x_{2ikl}(\lambda) : w_{-i}^{(k)} &\mapsto w_{-i}^{(k)} + \lambda w_0^{(l)} + \lambda^2 Q(w_0^{(l)}) x_i^{(k)}, \\ x_0^{(l)} &\mapsto x_0^{(l)} + \lambda x_i^{(k)} \end{aligned}$$

$$\begin{aligned} x_{3ikl}(\lambda) : x_{-i}^{(k)} &\mapsto x_{-i}^{(k)} + \lambda x_0^{(l)} + \lambda^2 Q(x_0^{(l)}) w_i^{(k)}, \\ w_0^{(l)} &\mapsto w_0^{(l)} + \lambda w_i^{(k)} \end{aligned}$$

$$\begin{aligned} x_{4ikl}(\lambda) : x_{-i}^{(k)} &\mapsto w_{-i}^{(k)} + \lambda w_0^{(l)} + \lambda^2 Q(w_0^{(l)}) w_i^{(k)}. \\ x_0^{(l)} &\mapsto x_0^{(l)} + \lambda w_i^{(k)} \end{aligned}$$

(2) For all $i, j \neq 0$ with $j < i$, all k, l, and every $\lambda \in \mathbb{F}_q$:

$$\begin{aligned} x_{5ijkl}(\lambda) : w_{-i}^{(k)} &\mapsto w_{-i}^{(k)} + \lambda w_{-j}^{(l)}, \\ x_j^{(l)} &\mapsto x_j^{(l)} + \lambda x_i^{(k)} \end{aligned}$$

$$\begin{aligned} x_{6ijkl}(\lambda) : w_{-i}^{(k)} &\mapsto w_{-i}^{(k)} + \lambda x_{-j}^{(l)}, \\ w_j^{(l)} &\mapsto w_j^{(l)} + \lambda x_i^{(k)} \end{aligned}$$

$$\begin{aligned} x_{7ijkl}(\lambda) : x_{-i}^{(k)} &\mapsto x_{-i}^{(k)} + \lambda w_{-j}^{(l)}. \\ x_j^{(l)} &\mapsto x_j^{(l)} + \lambda w_i^{(k)} \end{aligned}$$

(3) For all $i > 1$, $j < i$ with $j \neq \pm 1$, all k, l, and every $\lambda \in \mathbb{F}_q$:

$$\begin{aligned} x_{8ijkl}(\lambda) : v_{-i}^{(k)} &\mapsto v_{-i}^{(k)} + \lambda v_{-j}^{(l)}. \\ v_j^{(l)} &\mapsto v_j^{(l)} + \lambda v_i^{(k)} \end{aligned}$$

(4) For all $j < i$ with $j \neq \pm 1$, all k, l, and every $\lambda \in \mathbb{F}_q$:

$$\begin{aligned} x_{9ijkl}(\lambda) : \quad w_{-i}^{(k)} &\mapsto w_{-i}^{(k)} + \lambda v_{-j}^{(l)}, \\ v_j^{(l)} &\mapsto v_j^{(l)} + \lambda x_i^{(k)} \\ x_{10,ijkl}(\lambda) : x_{-i}^{(k)} &\mapsto x_{-i}^{(k)} + \lambda v_{-j}^{(l)}. \\ v_j^{(l)} &\mapsto v_j^{(l)} + \lambda w_i^{(k)} \end{aligned}$$

(5) For all $i \geq 1$, all k, l, and every $\lambda \in \mathbb{F}_q$:

$$\begin{aligned} x_{11,ikl}(\lambda) : w_{-i}^{(k)} &\mapsto w_{-i}^{(k)} + \lambda v_1^{(l)} + \lambda^2 x_i^{(k)}, \\ v_{-1}^{(l)} &\mapsto v_{-1}^{(l)} + \lambda x_i^{(k)} \end{aligned}$$

$$\begin{aligned} x_{12,ikl}(\lambda) : x_{-i}^{(k)} &\mapsto x_{-i}^{(k)} + \lambda v_1^{(l)} + \lambda^2 w_i^{(k)}. \\ v_{-1}^{(l)} &\mapsto v_{-1}^{(l)} + \lambda w_i^{(k)} \end{aligned}$$

(6) For all $i > 1$, all k, l, and every $\lambda, \mu \in \mathbb{F}_q$:

$$\begin{aligned} x_{13,ikl}(\lambda, \mu) : v_{-i}^{(k)} &\mapsto v_{-i}^{(k)} + \lambda w_0^{(l)} + \mu x_0^{(l)} + (\lambda\mu + (\lambda^2 + \mu^2) Q(w_0^{(l)})) v_i^{(k)}, \\ w_0^{(l)} &\mapsto w_0^{(l)} + \mu v_i^{(k)} \\ x_0^{(l)} &\mapsto x_0^{(l)} + \lambda v_i^{(k)} \end{aligned}$$

$$\begin{aligned} x_{14,ikl}(\lambda) : \quad v_{-i}^{(k)} &\mapsto v_{-i}^{(k)} + \lambda v_i^{(l)}. \\ v_{-i}^{(l)} &\mapsto v_{-i}^{(l)} + \lambda v_i^{(k)} \end{aligned}$$

(7) For all $i > 1$, all k, l, and every $\lambda, \mu \in \mathbb{F}_q$:

$$\begin{aligned} x_{15,ikl}(\lambda,\mu): v_{-i}^{(k)} &\mapsto v_{-i}^{(k)} + \lambda v_{-1}^{(l)} + \mu v_1^{(l)} + (\lambda\mu + \mu^2 + \lambda^2 Q(v_{-1}^{(l)}))v_i^{(k)}. \\ v_{-1}^{(l)} &\mapsto v_{-1}^{(l)} + \mu v_i^{(k)} \\ v_1^{(l)} &\mapsto v_1^{(l)} + \lambda v_i^{(k)} \end{aligned}$$

(8) For all k, l, and every $\lambda, \mu \in \mathbb{F}_q$ satisfying $\mu^2 = \lambda^2 + \lambda$:

$$\begin{aligned} x_{16,kl}(\lambda): v_{-1}^{(k)} &\mapsto v_{-1}^{(k)} + \lambda v_1^{(k)} + \mu v_1^{(l)}. \\ v_{-1}^{(l)} &\mapsto v_{-1}^{(l)} + \mu v_1^{(k)} + \lambda v_1^{(l)} \end{aligned}$$

(9) For all k, l, m, and every $\lambda, \mu, \gamma, \delta \in \mathbb{F}_q$ satisfying $\delta^2 = \lambda\mu + \gamma^2 + \gamma + (\lambda^2 + \mu^2)Q(w_0^{(l)})$:

$$\begin{aligned} x_{17,klm}(\lambda,\mu,\gamma): v_{-1}^{(k)} &\mapsto v_{-1}^{(k)} + \lambda w_0^{(l)} + \mu x_0^{(l)} + \gamma v_1^{(k)} + \delta v_1^{(m)}. \\ v_{-1}^{(m)} &\mapsto v_{-1}^{(m)} + \lambda w_0^{(l)} + \mu x_0^{(l)} + \delta v_1^{(k)} + \gamma v_1^{(m)} \\ w_0^{(l)} &\mapsto w_0^{(l)} + \mu v_1^{(k)} + \mu v_1^{(m)} \\ x_0^{(l)} &\mapsto x_0^{(l)} + \lambda v_1^{(k)} + \lambda v_1^{(m)} \end{aligned}$$

(10) For all k, l, m:

$$\begin{aligned} x_{18,klm}: v_{-1}^{(k)} &\mapsto v_{-1}^{(k)} + v_1^{(l)} + v_1^{(m)}. \\ v_{-1}^{(l)} &\mapsto v_{-1}^{(l)} + v_1^{(k)} + v_1^{(m)} \\ v_{-1}^{(m)} &\mapsto v_{-1}^{(m)} + v_1^{(k)} + v_1^{(l)} \end{aligned}$$

5.2.2 *Factors* $\mathbf{Sp}_{2a_i}(q)$

These factors arise from a summand $W(m_i)^{[a_i]}$ in (5.2.1) with m_i even. Let $u = W(2l)^{[a]}$, with basis labelled

$$w_{-(2l-1)}^{(1)}, x_{-(2l-1)}^{(1)}, \ldots, w_{-(2l-1)}^{(a)}, x_{-(2l-1)}^{(a)}, \ldots, w_{2l-1}^{(1)}, x_{2l-1}^{(1)}, \ldots, w_{2l-1}^{(a)}, x_{2l-1}^{(a)}.$$

Relative to this basis, the symplectic form has matrix

$$F = \begin{pmatrix} & & & K \\ & & K & \\ & \iddots & & \\ K & & & \end{pmatrix}$$

where $K = J^{[a]}$ (the block diagonal sum of a copies of $J = \begin{pmatrix} & 1 \\ 1 & \end{pmatrix}$).

Let $P = QL$ be the associated parabolic subgroup of $G = \mathrm{Sp}_{4al}(q)$ or $\Omega^+_{4al}(q)$. Elements of $C_P(u)$ project to elements of L having shape $x^{[l]} = x \oplus \cdots \oplus x$, where $x \in \mathrm{Sp}_{2a}(q)$ preserves the symplectic form with matrix K. Choose standard generators x_i $(1 \le i \le r)$ for this $\mathrm{Sp}_{2a}(q)$, and write $X_i = x_i^{[l]}$. For each i, we compute $v_i \in Q$ such that $u^{X_i} = u^{v_i}$. (This computation is carried out in the isomorphic copy of Q described by the power-conjugate presentation.) Then $X_i v_i^{-1} \in C_P(u)$, and

$$\langle X_i v_i^{-1} : 1 \le i \le r \rangle \le C_P(u)$$

maps onto a subgroup $\mathrm{Sp}_{2a}(q)$ of the Levi subgroup L. This gives us the factor $\mathrm{Sp}_{2a}(q)$ of $C_G(u)$.

5.2.3 *Factors* $I_{2a_i}(q)$ *for* $\mathrm{Sp}(V)$

Let $G = \mathrm{Sp}(V)$ and let u be as in (5.2.1). Here we construct the factors $I_{2a_i}(q) = \mathrm{Sp}_{2a_i}(q)$ or $\mathrm{O}^{\epsilon_i}_{2a_i}(q)$ of $C_G(u)$, given by (5.2.3).

First, if $m_i = 1$ for some i, then there is a summand $W(1)^{[a_i]}$ in (5.2.1), and clearly $C_G(u)$ has a subgroup $\mathrm{Sp}_{2a_i}(q)$ acting on this summand (and trivially on its orthogonal complement). We choose standard generators to generate this subgroup.

The other case where there is a factor $I_{2a_i}(q) = \mathrm{Sp}_{2a_i}(q)$ occurs when (5.2.1) has a summand $W(m_i)^{[a_i]} \oplus V_\beta(2k_j)$, where m_i is odd and $m_i = 2k_j \pm 1$. To simplify notation, denote the summand by $W(2l+1)^{[a]} \oplus V_\beta(2k)$, where $2l+1 = 2k \pm 1$. This is handled as in Sect. 5.2.2: we choose standard generators x_i $(1 \le i \le r)$ of $\mathrm{Sp}_{2a}(q)$ preserving the symplectic form with matrix K, and let $X_i = x_i^{[2l+1]}$ act on $W(2l+1)^{[a]}$ and trivially on $V_\beta(2k)$. For each i, compute $v_i \in Q$ such that $u^{X_i} = u^{v_i}$. Then $\langle X_i v_i^{-1} : 1 \le i \le r\rangle$ is a subgroup of $C_P(u)$ that maps onto $\mathrm{Sp}_{2a}(q)$, as required.

Finally, we deal with the case where the factor $I_{2a_i}(q)$ is $\mathrm{O}^{\epsilon_i}_{2a_i}(q)$. This occurs when (5.2.1) has a summand $W_{\beta_i,a_i}(m_i)$ with m_i odd, but no summand $V_\beta(2k_j)$ such that $m_i = 2k_j \pm 1$. Write the summand as $u = W_\beta(2l+1) \oplus W(2l+1)^{[a-1]}$.

In this case, elements of $C_P(u)$ project to elements of L having shape $x^{[2l+1]}$, where $x \in \mathrm{O}^{\epsilon}_{2a}(q)$ preserves the quadratic form with matrix

$$S = \begin{pmatrix} \beta & 1 & & & & & \\ 0 & \beta & & & & & \\ & & 0 & 1 & & & \\ & & 0 & 0 & & & \\ & & & & \ddots & & \\ & & & & & 0 & 1 \\ & & & & & 0 & 0 \end{pmatrix}. \tag{5.2.4}$$

So we choose standard generators x_i $(1 \le i \le r)$ of $\mathrm{O}^{\epsilon}_{2a}(q)$, and let $X_i = x_i^{[2l+1]}$ act on $W_\beta(2l+1) \oplus W(2l+1)^{[a-1]}$. For each i, compute $v_i \in Q$ such that $u^{X_i} = u^{v_i}$. Then $\langle X_i v_i^{-1} : 1 \le i \le r \rangle$ is a subgroup of $C_P(u)$ that maps onto $\mathrm{O}^{\epsilon}_{2a}(q)$, as required.

5.2.4 Factors $I_{2a_i}(q)$ for $\Omega(V)$

Let $G = \Omega(V)$ and let u be as in (5.2.1). Here we construct the factors $I_{2a_i}(q) = \mathrm{Sp}_{2a_i}(q)$ or $\mathrm{O}^{\epsilon_i}_{2a_i}(q)$ of $C_G(u)$, given by (5.2.3).

The orthogonal factors occur when (5.2.1) has a summand $W_{\beta_i,a_i}(m_i)$ with m_i odd, but no summand $V_\beta(2k_j)$ such that $m_i = 2k_j \pm 1$. Write the summand as $u = W_\beta(2l+1) \oplus W(2l+1)^{[a-1]}$. We produce these factors as in the previous section, with a small tweak. Let $\mathrm{O}^{\epsilon}_{2a}(q)$ be the orthogonal group preserving the quadratic form with matrix S as in (5.2.4). Choose standard generators x_i $(1 \le i \le s)$ of $\Omega^{\epsilon}_{2a}(q)$, and a reflection $r \in \mathrm{O}^{\epsilon}_{2a}(q) \setminus \Omega^{\epsilon}_{2a}(q)$. Let $X_i = x_i^{[2l+1]}$ and $Y = r^{[2l+1]}$ act on $W_\beta(2l+1) \oplus W(2l+1)^{[a-1]}$. Now compute $v_i, w \in Q$ such that $u^{X_i} = u^{v_i}$ and $u^Y = u^w$. Then $\langle X_i v_i^{-1} : 1 \le i \le s \rangle$ is a subgroup of $C_P(u)$ that maps onto $\Omega^{\epsilon}_{2a}(q)$, and Yw^{-1} is an element of $C_{\mathrm{O}(V)}(u) \setminus C_{\Omega(V)}(u)$ that maps to r. Write $r' = Yw^{-1}$, and let $r'_1, \dots, r'_t$ be the collection of such elements, one for each orthogonal factor in (5.2.3). (We use these elements in Sect. 5.2.5.)

The symplectic factors $I_{2a_i}(q) = \mathrm{Sp}_{2a_i}(q)$ occur when there is a summand $W(m_i)^{[a_i]} \oplus V_\beta(2k_j)$, where $m_i = 2k_j \pm 1$. Since $G = \Omega(V)$, there must be another summand $V_\gamma(2k')$. So let $u = W(2l+1)^{[a]} \oplus V_\beta(2k) \oplus V_\gamma(2k')$, where $2l+1 = 2k \pm 1$. First, as in the previous paragraph, we produce a subgroup X of $C_P(u)$ acting trivially on $V_\beta(2k) \oplus V_\gamma(2k')$ such that the action of X on $W(2l+1)^{[a]}$ maps onto $\Omega^{\epsilon}_{2a}(q)$. Secondly, we produce two additional elements of $C_P(u)$ as follows. With the usual labelling of basis vectors, let $w_i^{(j)}, x_i^{(j)}$ $(-2l \le i \le 2l$, $1 \le j \le a)$ be the basis of $W(2l+1)^{[a]}$, and label the bases of the two V-blocks as

$$\begin{aligned} &v_{-(2k-1)}, \ldots, v_{2k-1} \text{ for } V_\beta(2k), \\ &v'_{-(2k'-1)}, \ldots, v'_{2k'-1} \text{ for } V_\gamma(2k'). \end{aligned}$$

Now define the following linear maps which fix all basis vectors other than those listed:

$$\begin{aligned} s_1 : w_0^{(1)} &\mapsto w_0^{(1)} + x_0^{(1)} + v'_1, \\ v'_{-1} &\mapsto v'_{-1} + x_0^{(1)}, \\ w_i^{(1)} &\mapsto w_i^{(1)} + x_i^{(1)} \text{ for } i \ge 1 \end{aligned}$$

$$\begin{aligned} s_2 : x_0^{(1)} &\mapsto x_0^{(1)} + w_0^{(1)} + v_1, \\ v_{-1} &\mapsto v_{-1} + w_0^{(1)}, \\ x_i^{(1)} &\mapsto x_i^{(1)} + w_i^{(1)} \text{ for } i \ge 1. \end{aligned}$$

Observe that $s_1, s_2 \in \Omega(V)$ and $\langle X, s_1, s_2\rangle$ maps onto $\mathrm{Sp}_{2a}(q)$ in the Levi subgroup L. So if we compute $y_i \in Q$ such that $u^{s_i} = u^{y_i}$ for $i = 1, 2$, then $\langle X,\ s_1y_1^{-1},\ s_2y_2^{-1}\rangle$ is a subgroup of $C_P(u)$ mapping onto $\mathrm{Sp}_{2a}(q)$, as required.

5.2.5 The Factor $Z_2^{t+\delta}$

Let u be as in (5.2.1). For $G = \mathrm{Sp}(V)$, we cover the factor $Z_2^{t+\delta}$ simply by adding as generators of $C_G(u)$ each V-block in (5.2.1), acting trivially on the orthogonal complement. Let $v_1, \ldots, v_s$ be these generators. For $G = \Omega(V)$ these elements lie in $\mathrm{O}(V) \setminus G$; so we add as generators of $C_G(u)$ the pairwise products among $v_1, \ldots, v_s, r'_1, \ldots, r'_t$, where the r'_i are the generators defined in Sect. 5.2.4. Let S be the subgroup generated by these generators. Note that to construct the centralizer of u in $\mathrm{O}(V)$, we include all of the generators $v_1, \ldots, v_s, r'_1, \ldots, r'_t$.

5.2.6 Unipotent Radical R

Let u be as in (5.2.1). We now compute the unipotent radical R of $C_G(u)$; it has order q^{M_u} as in Theorem 5.2.1(i). We already computed $C_Q(u)$ in Sect. 5.2.1. By Conjecture 5.2.3, which we assume, $|RC_Q(u)/C_Q(u)| = q^t$, where $J = \{j : b_j = 2\}$ and $t = \sum_{j\in J}(2a_j + 1)$. Each summand of (5.2.1) of the form $W(2k_j)^{[a_j]} \oplus V_{\beta_j,2}(2k_j)$ contributes q^{2a_j+1} to this order. We let

$$u = W(2k)^{[a]} \oplus V(2k) \oplus V_\beta(2k), \tag{5.2.5}$$

and aim to construct a subgroup R of $C_G(u)$ such that $|RC_Q(u)/C_Q(u)| = q^{2a+1}$.

Symplectic Case

Let $G = \mathrm{Sp}(V)$, and let u be as in (5.2.5). We compute R in four steps.

1. In Sect. 5.2.2 we constructed a subgroup, say X, of $C_P(u)$ acting on the summand $W(2k)^a$ that maps onto $\mathrm{Sp}_{2a}(q)$.
2. Consider the summand $U = V(2k) \oplus V_\beta(2k)$ of (5.2.5), and let $u_0 = u \downarrow U$. We work in $\mathrm{Sp}(U)$. Our basis of U is

$$v^{(1)}_{-(2k-1)}, v^{(2)}_{-(2k-1)}, \ldots, v^{(1)}_{2k-1}, v^{(2)}_{2k-1}, \tag{5.2.6}$$

with respect to which the symplectic form has matrix

$$\begin{pmatrix} & & I_2 \\ & \iddots & \\ I_2 & & \end{pmatrix}. \tag{5.2.7}$$

If $A^{[2k]} = A \oplus \cdots \oplus A$ preserves the form (where A is 2×2), then A lies in Y_0, where

$$Y_0 = \{A \in \mathrm{GL}_2(q) : AA^{\mathrm{t}} = I\} = \left\{ \begin{pmatrix} a+1 & a \\ a & a+1 \end{pmatrix} : a \in \mathbb{F}_q \right\} \cong \mathbb{F}_q^+.$$

Let

$$M_a = \begin{pmatrix} a+1 & a \\ a & a+1 \end{pmatrix}.$$

For $a \in \mathbb{F}_q$, let $l_a = M_a^{[2k]}$, and compute $y_a \in Q$ such that $u_0^{l_a} = u_0^{y_a}$. Now define

$$Y = \langle l_a y_a^{-1} : a \in \mathbb{F}_q \rangle,$$

a subgroup of $C_P(u_0)$ that maps onto a subgroup of L isomorphic to $\mathbb{F}_q^+$.
3. Suppose $a \geq 1$, and consider a summand $U_1 = W(2k) \oplus V(2k) \oplus V_\beta(2k)$ of (5.2.5); let $u_1 = u \downarrow U_1$. We work in $\mathrm{Sp}(U_1)$. Our basis of U_1 is

$$w^{(1)}_{-(2k-1)}, x^{(1)}_{-(2k-1)} v^{(2)}_{-(2k-1)}, v^{(3)}_{-(2k-1)}, \ldots, w^{(1)}_{2k-1}, x^{(1)}_{2k-1} v^{(2)}_{2k-1}, v^{(3)}_{2k-1}, \tag{5.2.8}$$

with respect to which the symplectic form has matrix

$$\begin{pmatrix} & & M \\ & \cdot^{\cdot^{\cdot}} & \\ M & & \end{pmatrix}, \text{ where } M = \begin{pmatrix} 0 & 1 & & \\ 1 & 0 & & \\ & & 1 & 0 \\ & & 0 & 1 \end{pmatrix}.$$

If $A^{[2k]}$ preserves the form (where A is 4×4), then $AMA^{\mathrm{t}} = M$. One such matrix is

$$A = \begin{pmatrix} 1 & 0 & 1 & 1 \\ 0 & 1 & 1 & 1 \\ 1 & 1 & 1 & 0 \\ 1 & 1 & 0 & 1 \end{pmatrix}.$$

Let $l = A^{[2k]}$, and compute $v \in Q$ such that $u_1^l = u_1^v$. Let $l_0 = lv^{-1} \in C_P(u_1)$. Let X_0 be the subgroup of $C_P(u_1)$ acting on $W(2k)$ that maps onto $\mathrm{Sp}_2(q)$ (constructed in Step 1). Let Y_0 be the subgroup of $C_P(u_1)$ acting on $V(2k)+V_\beta(2k)$ that maps onto $\mathbb{F}_q^+$ (constructed in Step 2). Now $Z = \langle X_0, Y_0, l_0 \rangle$ is a subgroup of $C_P(u_1)$ that maps onto the following subgroup Z_0 of $\mathrm{GL}_4(q)$ in the Levi subgroup L:

$$Z_0 = \left\langle \begin{pmatrix} A & \\ & I \end{pmatrix}, \begin{pmatrix} I & \\ & M_a \end{pmatrix}, l_0 : A \in \mathrm{Sp}_2(q),\ a \in \mathbb{F}_q \right\rangle.$$

Observe that Z_0 has the form $[q^3].\mathrm{Sp}_2(q)$ (where $[q^3]$ denotes a group of order q^3). The subgroup of Z mapping to the normal subgroup of order q^3 gives the subgroup R for the case $a = 1$ (in which case $|RC_Q(u)/C_Q(u)| = q^3$).

4. Let u be as in (5.2.5). If $a = 0$, then we constructed R in Step 2. Assume that $a \geq 1$. Let Z be the subgroup constructed in Step 3, acting on a summand $W(2k) \oplus V(2k) \oplus V_\beta(2k)$ and trivially on the orthogonal complement, and let X be the subgroup in Step 1 that maps onto $\mathrm{Sp}_{2a}(q)$. Define

$$C = \langle X,\ Z \rangle \leq C_P(u).$$

Now C maps onto a subgroup $[q^{2a+1}].\mathrm{Sp}_{2a}(q)$ of L. Hence, adding the subgroup S of Sect. 5.2.5,

$$\langle C_Q(u),\ C,\ S \rangle = C_P(u).$$

This completes the construction of $C_G(u)$ when $G = \mathrm{Sp}(V)$.

Orthogonal Case

Let $G = \Omega(V)$, and let u be as in (5.2.5). We again compute R in four steps, but some details are different.

1. This step is identical to that for the symplectic case: in Sect. 5.2.2 we constructed a subgroup, say X, of $C_P(u)$ acting on the summand $W(2k)^a$ that maps onto $\mathrm{Sp}_{2a}(q)$.
2. We consider the summand $U = V(2k) \oplus V_\beta(2k)$ of (5.2.5), with basis as in (5.2.6), and let $u_0 = u \downarrow U$. We work in $\Omega(U)$, with quadratic form Q_β as given in Sect. 5.1: namely, Q_β takes values $1, 1, \beta, \beta$ on $v_1^{(1)}, v_1^{(2)}, v_{-1}^{(1)}, v_{-1}^{(2)}$ respectively and 0 on the other basis vectors, and the associated symplectic form is (5.2.7). When $\beta \neq 0$, the matrix $l_a = M_a^{[2k]}$ defined in Step 2 of the symplectic case does not preserve the quadratic form Q_β. We adjust it as follows: for $a \in \mathbb{F}_q$, define $l'_a \in \Omega(U)$ by
$$\begin{aligned} l'_a : v_i^{(1)} &\mapsto (a+1)v_i^{(1)} + av_i^{(2)} \text{ for all } i \neq -1 \\ v_i^{(2)} &\mapsto av_i^{(1)} + (a+1)v_i^{(2)} \text{ for all } i \neq -1 \\ v_{-1}^{(1)} &\mapsto (a+1)v_{-1}^{(1)} + av_{-1}^{(2)} + a^2\beta v_1^{(1)} + a^2\beta v_1^{(2)} \\ v_{-1}^{(2)} &\mapsto av_{-1}^{(1)} + (a+1)v_{-1}^{(2)} + a^2\beta v_1^{(1)} + a^2\beta v_1^{(2)}, \end{aligned}$$
and it fixes all other basis vectors. Compute $y_a \in Q$ such that $u_0^{l'_a} = u_0^{y_a}$. Now $Y = \langle l'_a y_a^{-1} : a \in \mathbb{F}_q \rangle$ is a subgroup of $C_P(u_0)$ that maps onto a subgroup of L isomorphic to $\mathbb{F}_q^+$.
3. We work in $\Omega(U_1)$, where $U_1 = W(2k) \oplus V(2k) \oplus V_\beta(2k)$ is a summand of (5.2.5), with basis as in (5.2.8). Let $u_1 = u \downarrow U_1$. Again, we need to tweak the element $l = A^{[2k]}$ defined in Step 3 of the symplectic case, as it does not preserve the quadratic form. Define $l' \in \Omega(U_1)$ by
$$l' = A^{[k-1]} \oplus B \oplus A^{[k-1]},$$
where A is the 4×4 matrix in the symplectic case, and B is the 8×8 matrix corresponding to the basis vectors of weights -1 and 1,
$$w_{-1}^{(1)}, x_{-1}^{(1)}, v_{-1}^{(2)}, v_{-1}^{(3)}, w_1^{(1)}, x_1^{(1)}, v_1^{(2)}, v_1^{(3)};$$

so

$$B = \begin{pmatrix} 1\,0\,1\,1\,0\,\beta & \\ 0\,1\,1\,1\,\beta\,0 & \\ 1\,1\,1\,0\,\beta\,\beta & \\ 1\,1\,0\,1\,\beta\,\beta & \\ & 1\,0\,1\,1 \\ & 0\,1\,1\,1 \\ & 1\,1\,1\,0 \\ & 1\,1\,0\,1 \end{pmatrix}.$$

Now compute $v \in Q$ such that $u_1^{l'} = u_1^v$, and let $l_0 = l'v^{-1} \in C_P(u_1)$. Using this element we construct the subgroup R for the case $a = 1$ in similar fashion to the corresponding step for the symplectic case.

4. This step is identical to that for the symplectic case.

This completes the construction of $C_G(u)$ when $u \in G = \Omega(V)$. To construct $C_{O(V)}(u)$, we include the extra generators indicated in Sect. 5.2.5.

Let $G = \mathrm{O}(V)$. If $u \in G \setminus \Omega(V)$, then we cannot write down directly $C_G(u)$. But we use our approach to reduce its construction to a smaller problem: since $u^2 \in \Omega(V)$, we construct $H := C_G(u^2)$; if H is sufficiently "small", then we can use a standard centralizer algorithm [17, Chapter 4] to construct $C_H(u)$.

5.3 The Conjugacy Problem

Let $G = \mathrm{Sp}(V)$ or $\mathrm{O}(V)$ in characteristic 2. In this section we address the conjugacy problem: given unipotent $g \in G$, find the canonical class representative as in (5.1.1) that is conjugate to g. Before presenting the algorithm to do this, we prove five necessary lemmas.

Lemma 5.3.1 *Let $u = V_\beta(2k)$ or $W_\beta(2l+1)$, as defined in Sect. 5.1, where $\beta \in \{0, \alpha\}$. For $r \geq 1$, define $F_r = V(1-u)^r$. The actions of u on the spaces $F_r/\mathrm{Rad}(F_r)$ and $F_r^\perp/\mathrm{Rad}(F_r^\perp)$ are as in Table 5.1.*

Table 5.1 Actions of u on spaces

u	r	$u^{F_r/\mathrm{Rad}(F_r)}$	$u^{F_r^\perp/\mathrm{Rad}(F_r^\perp)}$
$V_\beta(2k)$	$r \leq k$	$V_\beta(2k-2r)$	0
	$k < r \leq 2k$	0	$V_\beta(2r-2k)$
$W_\beta(2l+1)$	$r \leq l$	$W_\beta(2l+1-2r)$	0
	$l < r \leq 2l+1$	0	$W_\beta(2r-2l-1)$

In the table, the convention is that for $\mathrm{Sp}(V)$, the elements $W_\beta(1)$ and $V_\beta(2)$ are just $W(1)$ and $V(2)$, whatever the value of β.

Proof Let $u = V_\beta(2k)$, defined as in Sect. 5.1 relative to the basis $v_{-(2k-1)}, \dots, v_{2k-1}$. If $r \leq k$, then $F_r = V(1-u)^r$ has basis $v_{-(2k-2r-1)}, \dots, v_{2k-1}$, and $\mathrm{Rad}(F_r) = \langle v_{2k-2r+1}, \dots, v_{2k-1}\rangle$. Hence u acts on $F_r/\mathrm{Rad}(F_r)$ as $V_\beta(2k-2r)$, as claimed. The other cases are similar. □

Lemma 5.3.2 *Let* $u = W(2)^{[a]} \oplus V(2)^{[b]} \oplus V_\beta(2)^{[c]} \in \mathrm{Sp}(V)$ *or* $\mathrm{O}(V)$, *where* $b+c \leq 2$ *and* $\dim V = 4a+2b+2c$. *Let* $w_1, \dots, w_r$ *be a basis of* $V(u-1)$, *and, for each* i, *let* $v_i \in V$ *be such that* $v_i(u-1) = w_i$. *Then* $b+c = 0$ *if and only if* $(v_i, w_i) = 0$ *for all* i.

Proof If $b+c=0$, then $u = W(2)^{[a]}$, so $(v, v(u-1)) = 0$ for all $v \in V$.

For the converse, suppose $(v_i, w_i) = 0$ for all i. Assume for a contradiction that $b+c \geq 1$. We assume that $b+c=2$, the case $b+c=1$ being similar and easier. Let X be a non-degenerate 4-space on which u acts as $V(2)^{[b]} \oplus V_\beta(2)^{[c]}$. Then u acts as $W(2)^{[a]}$ on $X^\perp$, so $(v, v(u-1)) = 0$ for all $v \in X^\perp$. Choose a basis e_1, e_2, f_1, f_2 of X such that $u-1$ maps $e_1 \mapsto f_1 \mapsto 0$, $e_2 \mapsto f_2 \mapsto 0$. There are two vectors, say w_1 and w_2, in the basis of $V(u-1)$ whose projections to X are linearly independent. So we may write

$$w_1 = \alpha_1 f_1 + \alpha_2 f_2 + x, \quad w_2 = \beta_1 f_1 + \beta_2 f_2 + x',$$

where $x, x' \in X^\perp$ and $\alpha_1 f_1 + \alpha_2 f_2$, $\beta_1 f_1 + \beta_2 f_2$ are linearly independent. Since $v_i(u-1) = w_i$,

$$\begin{aligned} v_1 &= \alpha_1 e_1 + \alpha_2 e_2 + \gamma_1 f_1 + \gamma_2 f_2 + y, \\ v_2 &= \beta_1 e_1 + \beta_2 e_2 + \delta_1 f_1 + \delta_2 f_2 + y', \end{aligned}$$

where $y, y' \in X^\perp$ and $y(u-1) = x$, $y'(u-1) = x'$. Then $(y, x) = (y', x') = 0$, and hence

$$\begin{aligned} (v_1, w_1) = 0 &\Rightarrow \alpha_1^2 + \alpha_2^2 = 0 \Rightarrow \alpha_1 = \alpha_2, \\ (v_2, w_2) = 0 &\Rightarrow \beta_1^2 + \beta_2^2 = 0 \Rightarrow \beta_1 = \beta_2. \end{aligned}$$

This contradicts the linear independence of $\alpha_1 f_1 + \alpha_2 f_2$ and $\beta_1 f_1 + \beta_2 f_2$. □

Lemma 5.3.3 *Let* $u = V_\beta(4) \in \mathrm{Sp}(V)$ *or* $\mathrm{O}(V)$, *where* $\dim V = 4$. *Suppose* $v \in V$ *is such that* $v(u-1)^3 \neq 0$ *and* $(v(u-1), v(u-1)^2) \neq 0$. *Write* $v_i = v(u-1)^{i-1}$ *for* $1 \leq i \leq 4$, *and define*

$$\lambda = (v_2, v_3), \quad \mu = (v_1, v_2).$$

Then $\beta = \alpha$ *if and only if* $x^2 + x + \lambda^{-1}\mu$ *is irreducible over* $\mathbb{F}_q$.

Proof Relative to the basis v_1, v_2, v_3, v_4, the matrix of u is J_4 and the matrix of the form $(\ ,\ )$ is

$$\begin{pmatrix} 0 & \mu & \lambda & \lambda \\ \mu & 0 & \lambda & 0 \\ \lambda & \lambda & 0 & 0 \\ \lambda & 0 & 0 & 0 \end{pmatrix}.$$

We change the basis as follows: let $\delta \in \mathbb{F}_q$ with $\delta^2 = \lambda^{-1}$, and set

$$\begin{aligned} e_1 &= \delta v_1 + \delta v_2 + \delta^3 \mu v_3, \\ e_2 &= \delta v_2, \\ f_2 &= \delta v_3, \\ f_1 &= \delta v_4. \end{aligned}$$

Now, relative to the basis $B = [e_1, e_2, f_2, f_1]$,

$$[u]_B = \begin{pmatrix} 1 & 1 & 1 & \lambda^{-1}\mu \\ & 1 & 1 & 0 \\ & & 1 & 1 \\ & & & 1 \end{pmatrix}, \quad (\ ,\ )_B = \begin{pmatrix} & & & 1 \\ & & 1 & \\ & 1 & & \\ 1 & & & \end{pmatrix}.$$

The matrix of u is that of $V_{\lambda^{-1}\mu}(4)$, and the conclusion follows. □

Lemma 5.3.4 *Let* $u = W_\beta(3) \in \mathrm{Sp}(V)$ *or* $\mathrm{O}(V)$, *where* $\dim V = 6$. *Suppose* $v_1, w_1 \in V$ *are such that*

$$v_1, v_1(u-1), v_1(u-1)^2, w_1, w_1(u-1), w_1(u-1)^2$$

is a basis of V. *Let* $v_2 = v_1(u-1)$, $w_2 = w_1(u-1)$, *and set*

$$\gamma_1 = (v_2, w_2), \ \gamma_2 = (v_1, v_2), \ \gamma_3 = (w_1, w_2).$$

Then $\beta = \alpha$ *if and only if the quadratic* $x^2 + \gamma_1 x + \gamma_2\gamma_3$ *is irreducible over* $\mathbb{F}_q$.

Proof Recall the basis defining $W_\beta(3)$ given in Sect. 5.1, namely

$$w_{-2}, x_{-2}, w_0, x_0, w_2, x_2.$$

Rewrite this basis as $e_1, f_3, e_2, f_2, e_3, f_1$ respectively, so that $(e_i, f_j) = \delta_{ij}$. We claim first that

$$\beta = 0 \Leftrightarrow \exists\, v \in V \text{ such that } (v,\ v(u-1)) = 0 \text{ and } v(u-1)^2 \neq 0. \tag{5.3.1}$$

The left to right implication is clear: if $\beta = 0$, then $v = e_1$ has the required property. Suppose $\beta = \alpha$. Let $v \in V$ be such that $v(u-1)^2 \neq 0$. Writing $v = \sum \lambda_i e_i + \sum \mu_i f_i$, this means that $\lambda_1 \neq 0$ or $\mu_3 \neq 0$. We now compute that

$$(v,\ v(u-1)) = \lambda_1^2\alpha + \lambda_1\mu_3 + \mu_3^2\alpha.$$

Taking $\lambda_1 \neq 0$, this gives

$$(v,\ v(u-1)) = \lambda_1^2\alpha^{-1}(\alpha^2 + \mu_3\lambda_1^{-1}\alpha + (\mu_3\lambda_1^{-1}\alpha)^2),$$

which is nonzero since the quadratic $x^2+x+\alpha$ (hence also $x^2+x+\alpha^2$) is irreducible over $\mathbb{F}_q$. If $\lambda_1 = 0$, then also $(v,\ v(u-1)) \neq 0$. This proves (5.3.1).

Next we establish one of the implications in the lemma: namely

$$x^2 + \gamma_1 x + \gamma_2\gamma_3 \text{ irreducible over } \mathbb{F}_q \Rightarrow \beta = \alpha. \tag{5.3.2}$$

We prove the contrapositive. Assume $\beta = 0$, and let v_1, w_1 be as in the hypothesis of the lemma. Write

$$v_1 = \lambda_1 e_2 + \lambda_3 f_3 + w,\ \ w_1 = \mu_1 e_2 + \mu_3 f_3 + w',$$

where $w, w' \in \langle e_2, e_3, f_1, f_2\rangle$. As $\beta = 0$, observe that $u = W(3)$ and we compute that

$$\gamma_1 = \lambda_1\mu_3 + \lambda_3\mu_1,\ \ \gamma_2 = \lambda_1\lambda_3,\ \ \gamma_3 = \mu_1\mu_3.$$

This gives $(\lambda_3\mu_1)^2+\gamma_1(\lambda_3\mu_1)+\gamma_2\gamma_3 = 0$, and hence the quadratic $x^2+\gamma_1 x+\gamma_2\gamma_3$ is reducible over $\mathbb{F}_q$, proving (5.3.2).

Now we complete the proof of the lemma by showing

$$x^2 + \gamma_1 x + \gamma_2\gamma_3 \text{ reducible over } \mathbb{F}_q \Rightarrow \beta = 0. \tag{5.3.3}$$

Suppose that $x^2 + \gamma_1 x + \gamma_2\gamma_3$ is reducible. Then there exists $\lambda \in \mathbb{F}_q$ such that $(\lambda\gamma_3)^2 + \gamma_1(\lambda\gamma_3) + \gamma_2\gamma_3 = 0$. Hence

$$(v_1 + \lambda w_1,\ (v_1 + \lambda w_1)(u-1)) = \lambda^2\gamma_3 + \lambda\gamma_1 + \gamma_2 = 0,$$

and therefore $\beta = 0$ by (5.3.1). □

Lemma 5.3.5 *Let $G = \mathrm{Sp}(V)$ or $\mathrm{O}(V)$. If u and v are unipotent elements of G as in Table 5.2, then u and v are conjugate by an element of* $\mathrm{Sp}(V)$ *or* $\Omega(V)$, *respectively.*

In the table $k, l \geq 1$, and, as before, our convention is that for $G = \mathrm{Sp}(V)$, the elements $W_\beta(1)$ and $V_\beta(2)$ are just $W(1)$ and $V(2)$, whatever the value of β.

Table 5.2 Conjugate unipotent elements

u	v
$V(2k) \oplus V(2k)$	$V_\alpha(2k) \oplus V_\alpha(2k)$
$V(2k+2) \oplus V(2k)$	$V_\alpha(2k+2) \oplus V_\alpha(2k)$
$V_\alpha(2k+2) \oplus V(2k)$	$V(2k+2) \oplus V_\alpha(2k)$
$V(2k)^{[3]}$	$W(2k) \oplus V(2k)$
$V(2k)^{[2]} \oplus V_\alpha(2k)$	$W(2k) \oplus V_\alpha(2k)$
$W(2l+1) \oplus W(2l+1)$	$W_\alpha(2l+1) \oplus W_\alpha(2l+1)$
$W(2l+1) \oplus V(2l+2)$	$W_\alpha(2l+1) \oplus V_\alpha(2l+2)$
$W(2l+1) \oplus V_\alpha(2l+2)$	$W_\alpha(2l+1) \oplus V(2l+2)$
$W(2l+1) \oplus V(2l)$	$W_\alpha(2l+1) \oplus V_\alpha(2l)$
$W(2l+1) \oplus V_\alpha(2l)$	$W_\alpha(2l+1) \oplus V(2l)$

Proof For $G = \mathrm{Sp}(V)$, this is [14, Lemma 3.4]. For $G = \mathrm{O}(V)$ it follows from the same result, combined with [14, Thm. 3.1]. □

The Algorithm

We now describe our algorithm for conjugacy testing in $G = \mathrm{Sp}(V)$ or $\mathrm{O}(V)$. Let $g \in G$ be unipotent. We need to compute a representative u of the form (5.1.1) (satisfying conditions (i)–(v) listed after (5.1.1)) such that g is conjugate to u. In the orthogonal case, the main step is to handle conjugacy in $\mathrm{O}(V)$; we address conjugacy in $\Omega(V)$ below.

Step 1. Let r be the largest Jordan block size for g. From the rational canonical form [16, §6.7] of g, we obtain a homocyclic component, say U, of the restriction $V \downarrow g$ corresponding to the block size r. (A *homocyclic component* is the sum of all cyclic summands isomorphic to a given one.) Then U contains all the blocks of size r for g. Moreover, U is non-degenerate: suppose for a contradiction that $U \cap U^\perp \neq 0$. Since $U \cap U^\perp$ is g-invariant, there exists $0 \neq w \in U \cap U^\perp$ such that $w(g-1) = 0$. Then $w = x(g-1)^{r-1}$ for some $x \in U$. Hence for every $v \in V$ such that $v(g-1)^{r-1} = 0$,

$$(w, v) = (x(g-1)^{r-1}, v) = (x, v(g^{-1}-1)^{r-1}) = 0.$$

It follows that $w \in V^\perp$, a contradiction.

Step 2. Suppose r is odd, so $U \downarrow g = \bigoplus_i W_{\beta_i}(r)$, where each $\beta_i \in \{0, \alpha\}$. We determine the β_i as follows.

Compute $v, w \in U$ such that $(v, w(g-1)^{r-1}) \neq 0$. Now $U_0 = \langle vg^i, wg^i : 0 \leq i \leq r-1 \rangle$ is non-degenerate and $U_0 \downarrow g = W_\beta(r)$ for some $\beta \in \{0, \alpha\}$. If $r = 1$ and $G = \mathrm{O}(V)$, then $\beta = 0$ or α according as the sign of $\mathrm{O}(U_0)$ is $+$ or $-$, respectively; and if $r = 1$ and $G = \mathrm{Sp}(V)$, then $\beta = 0$. Now assume $r \geq 3$. Let $F = U_0(g-1)^{(r-3)/2}$. By Lemma 5.3.1, g acts on $F/\mathrm{Rad}(F)$ as $W_\beta(3)$. We can compute β using Lemma 5.3.4.

Next we work with $U \cap U_0^\perp$ and repeat this process. The outcome is a sequence of values β_i such that

$$U \downarrow g = \bigoplus_i W_{\beta_i}(r), \tag{5.3.4}$$

where, as always, for $G = \mathrm{Sp}(V)$ the element $W_\beta(1)$ is just $W(1)$, whatever the value of β.

Step 3. Suppose r is even, say $r = 2k$, so

$$U \downarrow g = W(2k)^{[a]} \oplus \bigoplus_{i=1}^{b} V_{\gamma_i}(2k),$$

where $b \leq 2$. We first use Lemma 5.3.2 to determine whether $b = 0$ or not, as follows. Let $F = U(g-1)^{k-1}$. By Lemma 5.3.1, g acts on $F/\mathrm{Rad}(F)$ as $W(2)^{[a]} \oplus \bigoplus_{i=1}^{b} V_{\gamma_i}(2)$. We can use Lemma 5.3.2 to determine whether $b = 0$. If $b = 0$, then $U \downarrow g = W(2k)^{[a]}$, and we proceed to Step 4. If $b > 0$, then we determine the γ_i as follows. We find $v \in U$ such that $(v, v(g-1)^{2k-1}) \neq 0$. Now $U_0 = \langle vg^i : 0 \leq i \leq 2k-1\rangle$ is non-degenerate, and $U_0 \downarrow g = V_\gamma(2k)$ for some $\gamma \in \{0, \alpha\}$. If $k = 1$ and $G = \mathrm{O}(V)$, then $\beta = 0$ or α according as the sign of $\mathrm{O}(U_0)$ is $+$ or $-$, respectively. Now assume $k \geq 2$. Let $F = U_0(g-1)^{k-2}$. By Lemma 5.3.1, g acts on $F/\mathrm{Rad}(F)$ as $V_\gamma(4)$. We can compute γ using Lemma 5.3.3.

Next we work with $U \cap U_0^\perp$ and repeat this process. The outcome is a sequence of values $\gamma_1, \ldots, \gamma_b$ such that

$$U \downarrow g = W(2k)^{[a]} \oplus \bigoplus_{i=1}^{b} V_{\gamma_i}(2k), \tag{5.3.5}$$

where, as always, for $G = \mathrm{Sp}(V)$ the element $V_\gamma(2)$ is just $V(2)$, whatever the value of γ.

Step 4. Having determined (5.3.4) or (5.3.5) for the largest homocyclic component, we now work in $U^\perp$, and repeat Steps 1–3 for the next largest component. Repeating this for all the components, we end up with values of $\beta_i, \gamma_i \in \{0, \alpha\}$ such that g is conjugate to

$$\bigoplus_{r \text{ odd}} \bigoplus_i W_{\beta_i}(r) \oplus \bigoplus_{r \text{ even}} \left(W(r)^{[a_r]} \oplus \bigoplus_i V_{\gamma_i}(r) \right). \tag{5.3.6}$$

Our final task is to use Lemma 5.3.5 to convert this description to a canonical representative as in (5.1.1). We do this in the next step.

Step 5. Let the Jordan form of g be $\bigoplus_r J_r^{[n_r]}$, and define the following sets:

$$T = \{r : n_r > 0\},$$
$$S = \{r \in T : r \text{ odd}\} \cup \{r \in T : r \text{ even and } \exists\, V_{\gamma_i}(r) \text{ in (5.3.6)}\}.$$

For $r \in S$ and $\beta \in \{0, \alpha\}$, define $R_\beta(r)$ as follows:

r odd: $R_\beta(r) = W_\beta(r) \oplus W(r)^{[\frac{1}{2}n_r - 1]}$;
r even: $R_\beta(r) = V_\beta(r) \oplus V(r)^{[b]} \oplus W(r)^{[c]}$, where $b \le 1, b + 2c = n_r - 1$.

(Thus the Jordan form of $R_\beta(r)$ is $J_r^{[n_r]}$ in both cases.)
For $r, s \in S$, write $r \sim s$ if either $|r - s| = 1$, or $|r - s| = 2$ and both r and s are even. Define $\sim$ to be the minimal equivalence relation on T extending this (so each element of $T \setminus S$ is its own equivalence class).

Let S_0 be an equivalence class in T, and let $n(S_0)$ be the total number of subscripts β_i or γ_i in (5.3.6) that equal α for block sizes in S_0. Define $\alpha(S_0) \in \{0, \alpha\}$ as follows:

$$G = \mathrm{O}(V):\ \alpha(S_0) = \begin{cases} 0, & \text{if } n(S_0) \text{ is even} \\ \alpha, & \text{if } n(S_0) \text{ is odd} \end{cases}$$

$$G = \mathrm{Sp}(V):\alpha(S_0) = \begin{cases} 0, & \text{if } n(S_0) \text{ is even, or if } 2 \in S_0 \\ \alpha, & \text{if } n(S_0) \text{ is odd and } 2 \notin S_0. \end{cases}$$

We now identify the canonical representative $R(S_0)$ for each equivalence class S_0:

(a) if $S_0 = \{t\}$ with $t \in T \setminus S$, then let $R(S_0) = W(t)^{[n_t/2]}$;
(b) if $S_0 = \{t\}$ with t odd, then let $R(S_0) = R_{\alpha(S_0)}(t)$;
(c) otherwise $S_0 \subseteq S$ and $\exists$ even $t \in S_0$: let $r_0 = \min\{r \in S_0 : r \text{ even}\}$ and set

$$R(S_0) = R_{\alpha(S_0)}(r_0) \oplus \bigoplus_{r \in S_0, r \neq r_0} R_0(r).$$

Theorem 5.3.6 *The canonical representative for the element g as in (5.3.6) is*

$$\bigoplus_{S_0} R(S_0),$$

where the sum is over all the equivalence classes S_0.

Proof By definition of the equivalence relation, $\bigoplus_{S_0} R(S_0)$ satisfies conditions (i)–(v) listed after (5.1.1). By Lemma 5.3.5, the sum of the blocks in (5.3.6) with block sizes in S_0 is conjugate to $R(S_0)$. Hence g is conjugate to $\bigoplus_{S_0} R(S_0)$. □

To complete this section, we address conjugacy in $\Omega(V)$ versus $\mathrm{O}(V)$. As mentioned in Sect. 5.1, the only unipotent classes in $\mathrm{O}(V)$ that split in $\Omega(V)$ are those of the form $\bigoplus W(m_i)^{[a_i]}$ with all m_i even; in such a case, a second class representative is obtained by replacing one summand $W(m_i)$ by $W(m_i)^t$ where t is a reflection. Denote these representatives by u and u^t respectively. Given $g \in \Omega(V)$ that is $\mathrm{O}(V)$-conjugate to u, we determine whether g is $\Omega(V)$-conjugate to u or u^t, as follows. Using the results of Sect. 5.4, compute $y \in \mathrm{O}(V)$ such that $g^y = u$. Then g is $\Omega(V)$-conjugate to u if and only if $y \in \Omega(V)$.

5.4 Constructing a Conjugating Element

In this section we solve the conjugation problem for classical groups G in bad characteristic: given unipotent $g \in G$ that is conjugate to a class representative u, find $y \in G$ such that $g^y = u$. Recall that the class representatives are given in (5.1.1). As in Sect. 4.4, we compute the conjugating element y "block-by-block", so the main task is to solve the problem when u is a single block $V_\beta(2k)$ or $W_\beta(m)$. We address the symplectic and orthogonal groups separately, although both methods are similar.

5.4.1 Symplectic Groups

Let $G = \mathrm{Sp}(V)$ in characteristic 2. The problem for each of the four types of blocks is challenging.

Case 1: $u = V(2k)$

Let $u = V(2k) \in G = \mathrm{Sp}(V) \cong \mathrm{Sp}_{2k}(q)$. For odd characteristic, in Sect. 4.4.1 we used the nilpotent element of the Lie algebra $sp(V)$ corresponding to u (via the Cayley map) to express the basis $v_{-(2k-1)}, \dots, v_{2k-1}$. In characteristic 2, there is no Cayley map, but we mimic this method. Define the following nilpotent elements of $\mathrm{End}(V)$:

$$f = 1 + u, \quad e = f + f^2 + \cdots + f^{2k-1}.$$

The basis $v_{-(2k-1)}, \dots, v_{2k-1}$ with respect to which $V(2k)$ is defined in Sect. 5.1 can be expressed as follows, where $v = v_{-(2k-1)}$:

$$v,\ ve,\ ve^2, \dots, ve^{k-1},\ ve^{k-1}f,\ ve^{k-1}f^2, \dots, ve^{k-1}f^k. \tag{5.4.1}$$

With respect to this basis, the symplectic form defining G has matrix

$$\begin{pmatrix} & & & 1 \\ & & 1 & \\ & \cdot^{\cdot^{\cdot}} & & \\ 1 & & & \end{pmatrix}. \tag{5.4.2}$$

Now suppose $g \in G$ is conjugate to u. We aim to compute $y \in G$ such that $g^y = u$. Let

$$f_0 = 1 + g, \quad e_0 = f_0 + f_0^2 + \cdots + f_0^{2k-1}.$$

We seek $w \in V$ with the property that

$$w, \; we_0, \; we_0^2, \ldots, we_0^{k-1}, \; we_0^{k-1} f_0, \; we_0^{k-1} f_0^2, \ldots, we_0^{k-1} f_0^k \tag{5.4.3}$$

is a basis of V, with respect to which the symplectic form also has matrix (5.4.2). Then the element y sending each vector in (5.4.1) to the corresponding vector in (5.4.3) will lie in G and conjugate g to u, as required.

To compute such a vector w, we adopt a similar strategy to that in Sect. 4.4.1. To simplify notation, denote the sequence of operators

$$1, \; e_0, \; e_0^2, \ldots, e_0^{k-1}, \; e_0^{k-1} f_0, \; e_0^{k-1} f_0^2, \ldots, e_0^{k-1} f_0^k$$

by the symbols

$$m_0, \; m_1, \ldots, m_{2k-1}.$$

Then the vectors in the sequence (5.4.3) are $wm_0, \ldots, wm_{2k-1}$, and the requirement that the form has matrix (5.4.2) can be expressed as

$$(wm_i, \, wm_j) = \begin{cases} 1, & \text{if } i + j = 2k - 1 \\ 0, & \text{otherwise.} \end{cases} \tag{5.4.4}$$

Choose $z \in V \setminus V(g - 1)$. We aim to solve the following equation for $w \in V$ and $a_i \in \mathbb{F}_q$:

$$z = w \left(a_0 + \sum_{i \text{ odd}, \, 1 \le i \le 2k-3} a_i m_i \right),$$

such that w satisfies (5.4.4). To do this, we compute the values

$$\alpha_i = (z, \, zm_i) \quad (i \text{ odd}, \, 1 \le i \le 2k - 1).$$

The requirement (5.4.4) gives k quadratic equations in the a_i. The two simplest equations are

$$a_0^2 = \alpha_{2k-1}, \; a_1^2 + a_0 a_1 = \alpha_{2k-3}.$$

These arise from evaluating α_{2k-1} and α_{2k-3} imposing (5.4.4), as follows:

$$\begin{aligned}\alpha_{2k-1} &= (z, \, ze_0^{k-1} f_0^k) \\ &= (a_0 w + a_1 w e_0 + \cdots, a_0 w e_0^{k-1} f_0^k + \cdots) \\ &= a_0^2\end{aligned}$$

$$\begin{aligned}\alpha_{2k-3} &= (z, \, ze_0^{k-1} f_0^{k-2}) \\ &= (a_0 w + a_1 w e_0 + \cdots, \; a_0 w e_0^{k-1} f_0^{k-2} + a_1 w e_0^{k-1} (f_0^{k-1} + f_0^k) + \cdots) \\ &= a_0 a_1 + a_1^2.\end{aligned}$$

Given a solution for a_0, a_1 to these equations, we can successively solve for a_3, $a_5, \ldots, a_{2k-3}$ the ensuing equations with left hand sides $\alpha_{2k-5}, a_{2k-7}, \ldots, \alpha_1$. For example, when $k = 4$, the equations are

$$\begin{aligned}\alpha_7 &= a_0^2, \\ \alpha_5 &= a_0 a_1 + a_1^2, \\ \alpha_3 &= a_3 a_0 + a_1^2 + a_0 a_1, \\ \alpha_1 &= a_5 a_0 + a_3^2 + a_1 a_3 + a_0 a_3.\end{aligned}$$

Having solved for the a_i, we let

$$M = a_0 I + \sum_{i \text{ odd}, \, 1 \le i \le 2k-3} a_i m_i,$$

and set $w = zM^{-1}$. Now w is a vector satisfying (5.4.4), as required.

Case 2: $u = V_\alpha(2k)$

Let $u = V_\alpha(2k) \in G = \mathrm{Sp}(V) \cong \mathrm{Sp}_{2k}(q)$. This is trickier than the $V(2k)$ case because we cannot list the basis $v_{-(2k-1)}, \ldots, v_{2k-1}$ in terms of two nilpotent elements, as in (5.4.1). We adjust the procedure for $V(2k)$ as follows. As before, define

$$f = 1 + u, \quad e = f + f^2 + \cdots + f^{2k-1}.$$

Writing $v = v_{-(2k-1)}$, the first part of the basis $v_{-(2k-1)}, \ldots, v_{-3}$ is $v, ve, \ldots, ve^{k-2}$. We compute an endomorphism h', a polynomial in f, such that

$$v_{-1} = v_{-3}(e + h').$$

Thus $v_{-1} = ve^{k-2}(e + h')$. Set $h = e^{k-2}h'$. Then the basis $v_{-(2k-1)}, \ldots, v_{2k-1}$ can be expressed as follows:

$$v,\ ve, \ldots, ve^{k-2},\ v(e^{k-1} + h),\ v(e^{k-1} + h)f, \ldots, v(e^{k-1} + h)f^k. \tag{5.4.5}$$

Note that $h = p(f)$ is a polynomial in f of degree $2k - 1$. Here are the polynomials $p(f)$ for the first few values of k:

k	$p(f)$
2	αf^3
3	$\alpha f^4 + \alpha f^5$
4	$\alpha f^5 + \alpha^2 f^7$
5	$\alpha f^6 + \alpha f^7 + (\alpha^2 + \alpha) f^8 + \alpha f^9$
6	$\alpha f^7 + (\alpha^2 + \alpha) f^9 + \alpha^2 f^{10} + \alpha^3 f^{11}$

The rest of the algorithm is similar to the previous case. Suppose $g \in G$ is conjugate to u. We aim to compute $y \in G$ such that $g^y = u$. Let

$$f_0 = 1 + g,\quad e_0 = f_0 + f_0^2 + \cdots + f_0^{2k-1},\quad h_0 = p(f_0),$$

where $p(x)$ is the polynomial such that $h = p(f)$. We seek $w \in V$ with the property that

$$w,\ we_0, \ldots, we_0^{k-2},\ w(e_0^{k-1} + h_0),\ w(e_0^{k-1} + h_0)f_0, \ldots, w(e_0^{k-1} + h_0)f_0^k \tag{5.4.6}$$

is a basis of V, with respect to which the symplectic form has matrix (5.4.2). Then the element y sending each vector in (5.4.5) to the corresponding vector in (5.4.6) will lie in G and conjugate g to u, as required.

Denote the sequence of operators $1, e_0, \ldots, (e_0^{k-1} + h_0)f_0^k$ in (5.4.6) by the symbols $m_0, \ldots, m_{2k-1}$, so that the requirement that the form has matrix (5.4.2) can again be expressed as (5.4.4). Choose $z \in V \setminus V(g - 1)$. We aim to solve the following equation for $w \in V$ and $a_i \in \mathbb{F}_q$:

$$z = w\left(a_0 + \sum_{i \text{ odd},\ 1 \le i \le 2k-3} a_i m_i\right), \tag{5.4.7}$$

such that w satisfies (5.4.4). To do this, we compute the values

$$\alpha_i = (z,\ zm_i) \quad (i \text{ odd},\ 1 \leq i \leq 2k-1).$$

The requirement (5.4.4) gives k quadratic equations in the a_i. As before, these can be solved for the a_i. We let

$$M = a_0 I + \sum_{i \text{ odd},\ 1 \leq i \leq 2k-3} a_i m_i,$$

and set $w = zM^{-1}$. Now w is a vector satisfying (5.4.4), as required.

Case 3: $u = W(m)$

Let $u = W(m) \in G = \mathrm{Sp}(V) \cong \mathrm{Sp}_{2m}(q)$, defined as in Sect. 5.1 with respect to a basis $w_{-(m-1)}, x_{-(m-1)}, \ldots, w_{m-1}, x_{m-1}$. In a similar manner to Case 1, we express this basis in terms of the vectors $w = w_{-(m-1)}$, $x = x_{-(m-1)}$ and the following nilpotent operators:

$$e = 1 + u, \quad f = e + e^2 + \cdots + e^{m-1}.$$

The basis $w_{-(m-1)}, x_{-(m-1)}, \ldots, w_{m-1}, x_{m-1}$ is

$$w,\ we,\ \ldots,\ we^{m-1},\ x,\ xf,\ \ldots,\ xf^{m-1}, \tag{5.4.8}$$

and the symplectic form takes values

$$(we^i,\ xf^j) = \begin{cases} 1, & \text{if } i + j = m - 1 \\ 0, & \text{otherwise} \end{cases} \tag{5.4.9}$$

and $(we^i,\ we^j) = (xf^i,\ xf^j) = 0$ for all i, j.

Now suppose $g \in G$ is conjugate to u. We aim to compute $y \in G$ such that $g^y = u$. Let

$$e_0 = 1 + g, \quad f_0 = e_0 + e_0^2 + \cdots + e_0^{m-1}.$$

We seek $w', x' \in V$ with the property that

$$w',\ w'e_0,\ \ldots,\ w'e_0^{m-1},\ x',\ x'f_0,\ \ldots,\ x'f_0^{m-1} \tag{5.4.10}$$

is a basis of V with symplectic form values

$$(w'e_0^i,\ x'f_0^j) = \begin{cases} 1, & \text{if } i+j=m-1 \\ 0, & \text{otherwise} \end{cases} \tag{5.4.11}$$

and $(w'e_0^i,\ w'e_0^j) = (x'f_0^i,\ x'f_0^j) = 0$ for all i, j. Then the element y sending each vector in (5.4.8) to the corresponding vector in (5.4.10) will lie in G and conjugate g to u, as required.

Here is our algorithm to find such vectors w', x'. First, find $z, t \in V$ such that $\langle Vf_0,\ z,\ t\rangle = V$ and $(z,\ tf_0^{m-1}) \neq 0$. Now compute

$$\begin{aligned} \alpha_i &= (z,\ ze_0^i)\ \ (1 \le i \le \lfloor (m-1)/2 \rfloor), \\ \beta_i &= (t,\ tf_0^i)\ \ (1 \le i \le \lfloor (m-1)/2 \rfloor), \\ \gamma_i &= (z,\ tf_0^j)\ (0 \le j \le m-1). \end{aligned} \tag{5.4.12}$$

We consider separately the cases where m is even and odd.

(3a) Let $m = 2k$. We aim to solve the following equations for $w', x' \in V$ and $b_i, c_i, d_i \in \mathbb{F}_q$:

$$\begin{aligned} z &= w' + x' \textstyle\sum_{i=0}^{k-2} b_i f_0^i, \\ t &= w' \textstyle\sum_{i=1}^{k-1} c_{2i-1} e_0^{2i-1} + x' \sum_{i=0}^{2k-1} d_i f_0^i, \end{aligned} \tag{5.4.13}$$

such that w', x' satisfy (5.4.11). The values $\alpha_i, \beta_i, \gamma_i$, combined with (5.4.11), give $4k-2$ quadratic equations in b_i, c_i, d_i, and it turns out that these have a unique solution for b_i, c_i, d_i. For example, when $k = 3$ these equations are:

$$\begin{aligned} \alpha_2 &= b_1, \\ \alpha_1 &= b_0 + b_1, \\ \beta_2 &= c_1 d_0, \\ \beta_1 &= c_1 d_0 + c_3 d_0 + c_1 d_1 + c_1 d_2, \\ \gamma_5 &= d_0, \\ \gamma_4 &= d_1 + b_0 c_1, \\ \gamma_3 &= d_2 + b_0 c_1 + d_1 c_1, \\ \gamma_2 &= d_3 + b_0 c_1 + b_0 c_3, \\ \gamma_1 &= d_4 + b_0 c_1 + b_0 c_3 + b_1 c_1 + b_1 c_3, \\ \gamma_0 &= d_5. \end{aligned}$$

Note that (in general) $d_0 = \gamma_{2k-1}$, which is nonzero by the choice of z and t. For the case $k = 3$ we can solve the above equations successively for $d_0, b_1, b_0, c_1, d_1, d_2, c_3, d_3, d_4$ and finally d_5. A similar procedure applies in the general case.

For this solution for b_i, c_i, d_i, define

$$B = \sum_{i=0}^{k-2} b_i f_0^i, \quad C = \sum_{i=1}^{k-1} c_{2i-1} e_0^{2i-1}, \quad D = \sum_{i=0}^{2k-1} d_i f_0^i.$$

By (5.4.13), $z = w' + x'B$ and $t = w'C + x'D$. Hence, letting $M = BC + D$, and noting that D is invertible since $d_0 \neq 0$, we deduce that

$$x' = (zC + t)M^{-1}, \quad w' = z + x'B.$$

These vectors w', x' satisfy (5.4.11), as required.

(3b). Let $m = 2k + 1$. We aim to solve the following slightly different equations for $w', x' \in V$ and $b_i, c_i, d_i \in \mathbb{F}_q$:

$$\begin{aligned} z &= w' + x' \textstyle\sum_{i=0}^{k-1} b_{2i} f_0^{2i}, \\ t &= w' \textstyle\sum_{i=0}^{k-1} c_{2i} e_0^{2i} + x' \sum_{i=0}^{2k} d_i f_0^i, \end{aligned} \tag{5.4.14}$$

such that w', x' satisfy (5.4.11). This time the values $\alpha_i, \beta_i, \gamma_i$, combined with (5.4.11), give $4k + 1$ quadratic equations in the $4k + 1$ variables b_i, c_i, d_i. However, these are not as straightforward to solve as in the m even case. We solve the equations when $k = 2$—a case which shows the necessary features of the general case. For $k = 2$ the equations are:

$$\begin{aligned} \alpha_2 &= b_0, \\ \alpha_1 &= b_0 + b_2, \\ \beta_2 &= c_0 d_0, \\ \beta_1 &= c_0 d_0 + c_2 d_0 + c_0 d_1 + c_0 d_2, \\ \gamma_4 &= d_0 + b_0 c_0, \\ \gamma_3 &= d_1 + b_0 c_0, \\ \gamma_2 &= d_2 + b_0 c_0 + b_0 c_2 + b_2 c_0, \\ \gamma_1 &= d_3 + b_0 c_0 + b_0 c_2 + b_2 c_0, \\ \gamma_0 &= d_4 + b_2 c_2. \end{aligned}$$

Note first that the 1st, 3rd and 5th equations in this list show that d_0 is a root of the quadratic $x^2 + \gamma_4 x + \alpha_2 \beta_2$. A tedious exercise shows that this quadratic is indeed reducible. The choice of z and t implies that $\gamma_4 \neq 0$, so we can take d_0 to be a nonzero root. Given this, we can solve the equations uniquely for the rest of the b_i, c_i, d_i.

In the general case, the same features persist for solving the $4k + 1$ quadratic equations in the b_i, c_i, d_i, which is carried out computationally using a Gröbner basis algorithm [1, Chap. 1]. A certain quadratic must be reducible and implies that the equations have a unique solution. For this solution, define

$$B = \sum_{i=0}^{k-1} b_{2i} f_0^{2i}, \quad C = \sum_{i=0}^{k-1} c_{2i} e_0^{2i}, \quad D = \sum_{i=0}^{2k} d_i f_0^i.$$

Letting $M = BC + D$, we deduce that

$$x' = (zC + t)M^{-1}, \quad w' = z + x'B.$$

These vectors w', x' satisfy (5.4.11), as required.

Case 4: $u = W_\alpha(2l + 1)$

Let $u = W_\alpha(2l + 1) \in G = \mathrm{Sp}(V) \cong \mathrm{Sp}_{4l+2}(q)$, defined as in Sect. 5.1 with respect to the basis $w_{-2l}, x_{-2l}, \ldots, w_{2l}, x_{2l}$. Let $e = 1 + u$. Note that $e^{2l+1} = 0$. It is particularly difficult to describe the basis in terms of the nilpotent operator e (as we did in (5.4.8) in the previous case). We let $w = w_{-2l}$, $x = x_{-2l}$, and compute polynomials $f_i(t), g_i(t), h_i(t), k_i(t) \in \mathbb{F}_q[t]$ for $0 \le i \le 2l$, each of degree at most $2l$, such that

$$w_{-2l+2i} = wf_i(e) + xg_i(e), \quad x_{-2l+2i} = wh_i(e) + xk_i(e). \tag{5.4.15}$$

Note that $f_0(t) = k_0(t) = 1$, $g_0(t) = h_0(t) = 0$. The symplectic form takes values

$$(wf_i(e) + xg_i(e),\ wh_j(e) + xk_j(e)) = \begin{cases} 1, & \text{if } i + j = 2l \\ 0, & \text{otherwise} \end{cases} \tag{5.4.16}$$

and $(wf_i(e)+xg_i(e),\ wf_j(e)+xg_j(e)) = (wh_i(e)+xk_i(e),\ wh_j(e)+xk_j(e)) = 0$ for all i, j.

For example, when $l = 1$, $u = W_\alpha(3)$, the basis $w_{-2}, w_0, w_2, x_{-2}, x_0, x_2$ is

$$w,\ we + \alpha xe^2,\ we^2,\ x,\ \alpha we^2 + x(e + e^2),\ xe^2,$$

and so we can identify the polynomials:

i	$f_i(t)$	$g_i(t)$	$h_i(t)$	$k_i(t)$
0	1	0	0	1
1	t	αt^2	αt^2	$t + t^2$
2	t^2	0	0	t^2

For general l, we construct the relevant polynomials f_i, g_i, h_i, k_i by machine.

Our strategy is the usual one. Suppose $g \in G$ is conjugate to u. We aim to compute $y \in G$ such that $g^y = u$. We let $e_0 = 1 + g$ and seek $w', x' \in V$ with the property that

$$w' f_i(e_0) + x' g_i(e_0), \quad w' h_i(e_0) + x' k_i(e_0) \quad (0 \leq i \leq 2l) \tag{5.4.17}$$

is a basis of V with symplectic form values

$$(w' f_i(e_0) + x' g_i(e_0),\ w' h_j(e_0) + x' k_j(e_0)) = \begin{cases} 1, & \text{if } i + j = 2l \\ 0, & \text{otherwise} \end{cases} \tag{5.4.18}$$

and

$$\begin{aligned} &(w' f_i(e_0) + x' g_i(e_0),\ w' f_j(e_0) + x' g_j(e_0)) \\ &\quad = (w' h_i(e_0) + x' k_i(e_0),\ w' h_j(e_0) + x' k_j(e_0)) = 0 \end{aligned}$$

for all i, j. Then the element y sending each vector in (5.4.15) to the corresponding vector in (5.4.17) will lie in G and conjugate g to u, as required.

Here is our algorithm to find such vectors w', x'. First, find $z, t \in V$ such that $\langle V e_0,\ z,\ t \rangle = V$, and compute the scalars

$$\begin{aligned} \alpha_i &= (z,\ z f_i(e_0) + t g_i(e_0)) \quad (1 \leq i \leq 2l), \\ \beta_i &= (t,\ z h_i(e_0) + t k_i(e_0)) \quad (1 \leq i \leq 2l), \\ \gamma_i &= (z,\ z h_i(e_0) + t k_i(e_0)) \quad (0 \leq i \leq 2l). \end{aligned} \tag{5.4.19}$$

Now we aim to solve the following equations for $w', x' \in V$ and $b_i, c_i, d_i \in \mathbb{F}_q$:

$$\begin{aligned} z &= w' + \textstyle\sum_{i=0}^{l-1} b_i (w' h_{2i}(e_0) + x' k_{2i}(e_0)), \\ t &= \textstyle\sum_{i=0}^{l-1} c_i (w' f_{2i}(e_0) + x' g_{2i}(e_0)) + \sum_{i=0}^{2l} d_i (w' h_i(e_0) + x' k_i(e_0)) \end{aligned} \tag{5.4.20}$$

such that w', x' satisfy (5.4.18). The values $\alpha_i, \beta_i, \gamma_i$, combined with (5.4.20) and (5.4.18), give $6l + 1$ quadratic equations in b_i, c_i, d_i. Generating these quadratics is more complicated than in previous cases, as we do not know the polynomials f_i, g_i, h_i, k_i explicitly. To generate the quadratic equations, we proceed as follows.

Let b_i $(0 \leq i \leq l - 1)$, c_i $(0 \leq i \leq l - 1)$, and d_i $(0 \leq i \leq 2l)$ be indeterminates, and let F be the field of rational functions over $\mathbb{F}_q$ in the $4l + 1$ indeterminates. Let W, X, E be indeterminates over F, and define the ring

$$R = \frac{F[W, X, E]}{(E^{2l+1})}.$$

Define the following elements of R for $0 \le i \le 2l$:

$$V_i = Wf_i(E) + Xg_i(E), \quad V_{2l+1+i} = Wh_i(E) + Xk_i(E). \tag{5.4.21}$$

(Strictly speaking, these should be defined as cosets of the ideal (E^{2l+1}), but since all the polynomials f_i, g_i, h_i, k_i have degree less than $2l+1$, there is no ambiguity.) For $r = \sum_{i=0}^{4l+1} r_i V_i$, and $s = \sum_{i=0}^{4l+1} s_i V_i \in R$, where each $r_i, s_i \in F$, set

$$IP(r, s) = \sum_{i=0}^{4l+1} r_i s_{4l+1-i} \in F. \tag{5.4.22}$$

Note that $IP(r, s)$ is a quadratic expression in the coefficients r_i, s_i. If we specialize down to $W \mapsto w$, $X \mapsto x$, $E \mapsto e$, then $V_0, \dots, V_{4l+1}$ become the original basis vectors $w_{-2l}, \dots, x_{2l}$, and for $r_i, s_i \in \mathbb{F}_q$, we see that $r, s \in V$ and $IP(r, s)$ is the value of the symplectic form (r, s).

Now define the following elements of R:

$$Z = W + \textstyle\sum_{i=0}^{l-1} b_i \left(Wh_{2i}(E) + Xk_{2i}(E)\right),$$
$$T = \textstyle\sum_{i=0}^{l-1} c_i \left(Wf_{2i}(E) + Xg_{2i}(E)\right) + \sum_{i=0}^{2l} d_i \left(Wh_i(E) + Xk_i(E)\right).$$

Let A_i, B_i, C_i be the following elements of F:

$$\begin{aligned} A_i(b, c, d) &= IP(Z,\ Zf_i(E) + Tg_i(E)) \quad (1 \le i \le 2l), \\ B_i(b, c, d) &= IP(T,\ Zh_i(E) + Tk_i(E)) \quad (1 \le i \le 2l), \\ C_i(b, c, d) &= IP(Z,\ Zh_i(E) + Tk_i(E)) \quad (0 \le i \le 2l). \end{aligned} \tag{5.4.23}$$

As noted above, each of $A_i, B_i, C_i \in F$ is a quadratic expression in the variables b_i, c_i, d_i. The quadratic equations in b_i, c_i, d_i arising from (5.4.20) combined with (5.4.18) and (5.4.19) are

$$\begin{aligned} A_i(b, c, d) &= \alpha_i \quad (1 \le i \le 2l), \\ B_i(b, c, d) &= \beta_i \quad (1 \le i \le 2l), \\ C_i(b, c, d) &= \gamma_i \quad (0 \le i \le 2l). \end{aligned} \tag{5.4.24}$$

As an illustration, we compute the expression $A_1(b, c, d)$ in the case where $l = 1$ (so $u = W_\alpha(3)$). Here

$$Z = W + b_0 X, \quad T = c_0 W + d_0 X + d_1(\alpha WE^2 + X(E + E^2)) + d_2 XE^2.$$

To compute $A_1 := IP(Z,\ Zf_1(E)+Tg_1(E))$, we need to express Z and $Zf_1(E)+Tg_1(E)$ in terms of $V_0,\ldots,V_5$: these expressions are

$$Z = W + b_0X = V_0 + b_0V_3,$$

$$\begin{aligned} Zf_1(E)+Tg_1(E) &= WE + b_0XE + \alpha c_0WE^2 + \alpha d_0XE^2 \\ &= V_1 + \alpha(b_0+c_0)V_2 + b_0V_4 + (b_0+\alpha d_0+\alpha)V_5. \end{aligned}$$

Hence

$$A_1 = IP(Z,\ Zf_1(E)+Tg_1(E)) = \alpha b_0(b_0+c_0) + b_0 + \alpha d_0 + \alpha.$$

Returning to the general case, we use a Gröbner basis algorithm to find a solution for $b_i, c_i, d_i \in \mathbb{F}_q$ of the quadratic equations in (5.4.24). It may fail to produce a solution—in which case we re-choose the vectors z and t and repeat the above steps. (In practice, we obtain a solution after a small number of attempts.)

Given a solution b_i, c_i, d_i to (5.4.24), define

$$\begin{aligned} f(e_0) &= 1 + \textstyle\sum_{i=0}^{l-1} b_ih_{2i}(e_0), \\ g(e_0) &= \textstyle\sum_{i=0}^{l-1} b_ik_{2i}(e_0), \\ r(e_0) &= \textstyle\sum_{i=0}^{l-1} c_if_{2i}(e_0) + \sum_{i=0}^{2l} d_ih_i(e_0), \\ s(e_0) &= \textstyle\sum_{i=0}^{l-1} c_ig_{2i}(e_0) + \sum_{i=0}^{2l} d_ik_i(e_0). \end{aligned}$$

Now, by (5.4.14),

$$z = w'f(e_0) + x'g(e_0), \quad t = w'r(e_0) + x's(e_0). \tag{5.4.25}$$

Note that $f(e_0)$ is invertible. Let $M = g(e_0)r(e_0) + f(e_0)s(e_0)$. If M is invertible, or equivalently, $b_0c_0 \neq d_0$, then the solution of (5.4.25) is

$$x' = (zr(e_0) + tf(e_0))M^{-1}, \quad w' = (z + x'g(e_0))f(e_0)^{-1}. \tag{5.4.26}$$

(In practice, we observe that the inequality $b_0c_0 \neq d_0$ always holds, but we could not prove this.) These vectors w', x' satisfy (5.4.17) and (5.4.18), as required.

General Case

Suppose $g \in G = \mathrm{Sp}(V)$ is conjugate to a unipotent class representative

$$u = \bigoplus_i W(m_i)^{[c_i]} \oplus \bigoplus_j V(2k_j)^{[d_j]} \oplus \bigoplus_r W_\alpha(m_r') \oplus \bigoplus_s V_\alpha(2k_s'), \tag{5.4.27}$$

satisfying conditions (i)–(v) listed after (5.1.1). We aim to compute an element of G that conjugates g to u "block-by-block", using the work in Cases 1–4 above. However, the procedure is more subtle than in previous cases.

Let $2k$ be an even block size in (5.4.27), and let the contribution to u from blocks of this size be

$$W(2k)^{[a]} \oplus V_\alpha(2k)^{[b]} \oplus V(2k)^{[c]}, \tag{5.4.28}$$

where $b \leq 1$ and $b + c \leq 2$. By Lemma 5.3.5, this element of $\mathrm{Sp}(V_{4ak+2bk+2ck})$ is conjugate to

$$\begin{cases} W(2k)^{[a]}, & \text{if } b+c=0, \\ V(2k)^{[2a+c]} \oplus V_\alpha(2k)^{[b]}, & \text{if } b+c>0. \end{cases} \tag{5.4.29}$$

In (5.4.27), we replace each summand (5.4.28) by a summand (5.4.29) to get a conjugate element u'. We do this because in a space of the form (5.4.28) with $b+c > 0$ it is difficult to construct a W-block, so instead we convert the decomposition of the space into V-blocks, which can be easily found.

The Algorithm

We now describe our algorithm that performs the following task: given $g, h \in G$ that are both conjugate to u' (hence also to u), find $x \in G$ such that $g^x = h$.

The first step is to compute orthogonal decompositions of $V \downarrow g$ and $V \downarrow h$ into V-blocks and W-blocks, as follows.

Assume first that the largest block size in u' is even, say $2k$. Suppose there is a block $V(2k)$ or $V_\alpha(2k)$ present. We find $v \in V$ such that $(v,\ v(g-1)^{2k-1}) \neq 0$, and let $X = \langle vg^i : 0 \leq i \leq 2k-1\rangle$. Then $X \downarrow g$ is a V-block of size $2k$, and we can determine whether it is $V(2k)$ or $V_\alpha(2k)$ using Lemma 5.3.3 (as in Step 3 of the algorithm in Sect. 5.3). Now we repeat this computation in $X^\perp$. Note that we need to match the V-blocks thus constructed with those in u', which may involve rechoosing the vectors v in the above process.

Suppose there is no block $V(2k)$ or $V_\alpha(2k)$ present in u'. We find $w, x \in V$ such that $(w,\ x(g-1)^{2k-1}) \neq 0$, and let $X = \langle wg^i, xg^i : 0 \leq i \leq 2k-1\rangle$. Now $X \downarrow g$ is $W(2k)$, and we repeat with $X^\perp$.

To conclude this step, assume now that the largest block size is odd, say $2l+1$. We find $w, x \in V$ such that $(w,\ x(g-1)^{2l}) \neq 0$, and let $X = \langle wg^i, xg^i : 0 \leq i \leq 2l\rangle$. Then $X \downarrow g$ is $W(2l+1)$ or $W_\alpha(2l+1)$; using Lemma 5.3.4, we determine which (as in Step 2 of Sect. 5.3). Now we repeat this computation in $X^\perp$. As above, we need to match the W-blocks thus constructed with those in u', which may involve rechoosing the vectors w, x in this process.

We carry out the same procedure for the element h.

Now we have orthogonal decompositions

$$V \downarrow g = \bigoplus_{i=1}^{r} X_i, \quad V \downarrow h = \bigoplus_{i=1}^{r} Y_i,$$

where, for each i, the actions g^{X_i} and h^{Y_i} are conjugate to a V-block or a W-block, and are conjugate to each other. Moreover, both decompositions match that of the element u'.

Let $g_i = g^{X_i}$ and $h_i = h^{Y_i}$. Hence $g = \mathrm{diag}(g_1, \ldots, g_r)$ and $h = \mathrm{diag}(h_1, \ldots, h_r)$ with respect to the union of standard bases of the spaces X_i, Y_i. Let $\dim X_i = 2n_i$. Using the work for individual blocks in Cases 1–4 above, we compute $x_i \in \mathrm{Sp}_{2n_i}(q)$ such that $g_i^{x_i} = h_i$ for $1 \le i \le r$. Now $x = \mathrm{diag}(x_1, \ldots, x_r)$ is an element of G that conjugates g to h.

5.4.2 Orthogonal Groups

Let $G = \mathrm{O}(V)$ in characteristic 2. As for symplectic groups, the major part of the work is to find conjugating elements for individual V- and W-blocks. We first consider together the blocks $V(2k)$ and $V_\alpha(2k)$.

Case 1: $u = V_\beta(2k)$
Let $u = V_\beta(2k) \in G = \mathrm{O}(V)$, where $\beta \in \{0, \alpha\}$ and $V = V_{2k}$. Recall that $G \cong \mathrm{O}_{2k}^+(q)$ for $\beta = 0$, and $G \cong \mathrm{O}_{2k}^-(q)$ for $\beta = \alpha$. The element u is defined as in Sect. 5.1 with respect to the basis

$$v_{-(2k-1)}, v_{-(2k-3)}, \ldots, v_{2k-1},$$

with symplectic form given by $(v_i, v_{-i}) = 1$ for all i, and quadratic form $Q_\beta(v_{-1}) = \beta$, $Q_\beta(v_1) = 1$, all other values (v_i, v_j) and $Q_\beta(v_i)$ being 0.

We proceed as in Case 2 of Sect. 5.4.1, with a few tweaks to take account of the quadratic form. Let $f = 1 + u$, $e = f + f^2 + \cdots + f^{2k-1}$, and compute $h' \in \mathrm{End}(V_{2k})$, a polynomial in f, such that $v_{-1} = v_{-3}(e + h')$. If we set $h = e^{k-2}h'$, then the basis $v_{-(2k-1)}, \ldots, v_{2k-1}$ can be expressed as

$$v,\ ve, \ldots, ve^{k-2},\ v(e^{k-1} + h),\ v(e^{k-1} + h)f, \ldots, v(e^{k-1} + h)f^k. \tag{5.4.30}$$

Note that $h = 0$ if $\beta = 0$; and if $\beta = \alpha$, then $h = p(f)$, where $p(f)$ is the same polynomial as in Case 2 of Sect. 5.4.1.

Suppose $g \in \mathrm{O}(V)$ is conjugate to u. We aim to compute $y \in \mathrm{O}(V)$ such that $g^y = u$. As in Sect. 5.4.1, we let $f_0 = 1+g$, $e_0 = f_0+f_0^2+\cdots+f_0^{2k-1}$, $h_0 = p(f_0)$, and seek $w \in V$ such that

$$w,\ we_0, \ldots, we_0^{k-2},\ w(e_0^{k-1} + h_0),\ w(e_0^{k-1} + h_0)f_0, \ldots, w(e_0^{k-1} + h_0)f_0^k \tag{5.4.31}$$

is a basis of V, with respect to which the symplectic form has matrix (5.4.2), and

$$Q_\beta(w(e_0^{k-1}+h_0))=\beta,\quad Q_\beta(w(e_0^{k-1}+h_0)f_0)=1,$$

the rest of the Q_β-values on the basis being 0.

Denote the sequence of operators $1, e_0, \dots, (e_0^{k-1}+h_0)f_0^k$ in (5.4.31) by the symbols $m_0, \dots, m_{2k-1}$. Then the requirements on w are (5.4.4) and the equations

$$Q_\beta(wm_k)=\beta,\ Q_\beta(wm_{k+1})=1,\ \text{and } Q(wm_i)=0 \text{ for } i\neq k,k+1. \tag{5.4.32}$$

Choose $z\in V\setminus V(g-1)$. We aim to solve the following equation for $w\in V$ and $a_i\in\mathbb{F}_q$:

$$z=w\left(a_0+\sum_{i \text{ odd},\ 1\leq i\leq 2k-1} a_i m_i\right), \tag{5.4.33}$$

such that w satisfies (5.4.4) and (5.4.32). There is a difference with the corresponding Eq. (5.4.7) in Sect. 5.4.1: the sum on the right hand side includes a term for $i=2k-1$; the variable a_{2k-1} is needed to allow enough freedom to solve the extra Eq. (5.4.32).

To find a solution to (5.4.33), we compute the values

$$\begin{aligned}&\alpha_i=(z,zm_i)\ \ (i \text{ odd}, 1\leq i\leq 2k-1), \text{ and}\\ &\gamma=Q_\beta(z).\end{aligned}$$

The requirements (5.4.4) and (5.4.32) give k quadratic equations in the a_i, together with the equation

$$\gamma=\begin{cases}a_{k-1}^2\beta+a_0a_{2k-1}, & \text{if } k \text{ is even},\\ a_k^2+a_0a_{2k-1}, & \text{if } k \text{ is odd}.\end{cases}$$

As in Sect. 5.4.1, these $k+1$ equations can be solved for the a_i. We let

$$M=a_0I+\sum_{i \text{ odd},\ 1\leq i\leq 2k-1} a_i m_i,$$

and set $w=zM^{-1}$. Then w is a vector satisfying (5.4.4) and (5.4.32), as required.

Case 2: $u=W(m)$

This case is similar to Case 3 of Sect. 5.4.1, with some tweaks to take account of the quadratic form. Let $u=W(m)\in G=\mathrm{O}(V)\cong \mathrm{O}_{2m}^+(q)$, defined as in Sect. 5.1 with respect to a basis $w_{-(m-1)}, x_{-(m-1)}, \dots, w_{m-1}, x_{m-1}$. Let $e=1+u$ and

$f = e + e^2 + \cdots + e^{m-1}$. The basis $w_{-(m-1)}, x_{-(m-1)}, \ldots, w_{m-1}, x_{m-1}$ is

$$w,\ we,\ \ldots,\ we^{m-1},\ x,\ xf, \ldots, xf^{m-1},$$

the symplectic form takes values as in (5.4.9), and the quadratic form Q preserved by G takes the value 0 on every basis vector.

Suppose $g \in G$ is conjugate to u. We aim to compute $y \in G$ such that $g^y = u$. We let $e_0 = 1+g$, $f_0 = e_0+e_0^2+\cdots+e_0^{m-1}$, and seek $w', x' \in V$ with the property that

$$w',\ w'e_0,\ \ldots,\ w'e_0^{m-1},\ x',\ x'f_0, \ldots, x'f_0^{m-1}, \tag{5.4.34}$$

is a basis of V, with symplectic form values as in (5.4.11), and Q-values 0 on every basis vector.

Here is our algorithm to find such vectors w', x'. First, find $z, t \in V$ such that $\langle Vf,\ z,\ t\rangle = V$ and $(z,\ tf_0^{m-1}) \neq 0$. Now compute

$$\begin{aligned} \alpha_i &= (z,\ ze_0^i)\ (1 \leq i \leq \lfloor (m-1)/2 \rfloor), \\ \beta_i &= (t,\ tf_0^i)\ (1 \leq i \leq \lfloor (m-1)/2 \rfloor), \\ \gamma_i &= (z,\ tf_0^j)\ (0 \leq j \leq m-1), \\ \delta_1 &= Q(z), \\ \delta_2 &= Q(t). \end{aligned} \tag{5.4.35}$$

(2a) Let $m = 2k$. We aim to solve the following equations for $w', x' \in V$ and $b_i, c_i, d_i \in \mathbb{F}_q$:

$$\begin{aligned} z &= w' + x'\left(\textstyle\sum_{i=0}^{k-2} b_i f_0^i + b_{2k-1} f_0^{2k-1}\right), \\ t &= w' \textstyle\sum_{i=1}^{k} c_{2i-1} e_0^{2i-1} + x' \sum_{i=0}^{2k-1} d_i f_0^i, \end{aligned} \tag{5.4.36}$$

such that w', x' give a basis as in (5.4.34) with symplectic form values as in (5.4.11) and Q-values 0. Note the extra terms in these equations in comparison to (5.4.13): these are the terms $b_{2k-1} f_0^{2k-1}$ in the first equation and $c_{2k-1} e_0^{2k-1}$ in the second. (The extra variables b_{2k-1}, c_{2k-1} are needed to take care of the Q-values.) The values $\alpha_i, \beta_i, \gamma_i, \delta_1, \delta_2$, combined with (5.4.11) and the Q-values condition, give $4k$ quadratic equations in the variables b_i, c_i, d_i, and it turns out that these have a unique solution for b_i, c_i, d_i. For this solution, define

$$B = \sum_{i=0}^{k-2} b_i f_0^i + b_{2k-1} f_0^{2k-1},\quad C = \sum_{i=1}^{k} c_{2i-1} e_0^{2i-1},\quad D = \sum_{i=0}^{2k-1} d_i f_0^i.$$

By (5.4.36), $z = w' + x'B$ and $t = w'C + x'D$, so letting $M = BC + D$,

$$x' = (zC + t)M^{-1}, \quad w' = z + x'B.$$

These are the required vectors w', x'.

(2b) Let $m = 2k + 1$. We aim to solve the following slightly different equations for $w', x' \in V$ and $b_i, c_i, d_i \in \mathbb{F}_q$:

$$\begin{aligned} z &= w' + x' \textstyle\sum_{i=0}^{k} b_{2i} f_0^{2i}, \\ t &= w' \textstyle\sum_{i=0}^{k} c_{2i} e_0^{2i} + x' \sum_{i=0}^{2k} d_i f_0^i, \end{aligned} \tag{5.4.37}$$

such that w', x' give a basis as in (5.4.34) with symplectic form values as in (5.4.11) and Q-values 0. (Note the extra terms in comparison to (5.4.14), namely $b_{2k} f_0^{2k}$ and $c_{2k} e_0^{2k}$.) This time the values $\alpha_i, \beta_i, \gamma_i, \delta_1, \delta_2$, combined with (5.4.11) and the Q-values condition, give $4k+3$ quadratic equations in the variables b_i, c_i, d_i. These can be solved using a Gröbner basis algorithm. For this solution, define

$$B = \sum_{i=0}^{k} b_{2i} f_0^{2i}, \quad C = \sum_{i=0}^{k} c_{2i} e_0^{2i}, \quad D = \sum_{i=0}^{2k} d_i f_0^i.$$

Letting $M = BC + D$, we deduce that

$$x' = (zC + t)M^{-1}, \quad w' = z + x'B.$$

These are the required vectors w', x'.

Case 3: $u = W_\alpha(2l + 1)$

Let $u = W_\alpha(2l + 1) \in G = \mathrm{O}(V) \cong \mathrm{O}^-_{4l+2}(q)$, defined as in Sect. 5.1 with respect to the basis $w_{-2l}, x_{-2l}, \ldots, w_{2l}, x_{2l}$. Let $e = 1 + u$. We let $w = w_{-2l}$, $x = x_{-2l}$, and compute polynomials $f_i(t), g_i(t), h_i(t), k_i(t) \in \mathbb{F}_q[t]$ for $0 \le i \le 2l$, each of degree at most $2l$, such that

$$w_{-2l+2i} = wf_i(e) + xg_i(e), \quad x_{-2l+2i} = wh_i(e) + xk_i(e) \quad \text{for } 0 \le i \le 2l. \tag{5.4.38}$$

The symplectic form takes values as in (5.4.16), and the quadratic form Q preserved by G takes values 0 on the basis, apart from $Q(w_{-2}) = Q(w_0) = Q(x_0) = \alpha$, which means that

$$Q(wf_{l-1}(e) + xg_{l-1}(e)) = Q(wf_l(e) + xg_l(e)) = Q(wh_l(e) + xk_l(e)) = \alpha. \tag{5.4.39}$$

Our method is as in Case 4 of Sect. 5.4.1. Suppose $g \in G$ is conjugate to u. We aim to compute $y \in G$ such that $g^y = u$. We let $e_0 = 1 + g$ and seek $w', x' \in V$ with the property that

$$w' f_i(e_0) + x' g_i(e_0), \quad w' h_i(e_0) + x' k_i(e_0) \quad (0 \le i \le 2l) \tag{5.4.40}$$

is a basis of V with symplectic form values as in (5.4.18), and Q-values 0 apart from

$$\begin{aligned} Q(w' f_{l-1}(e_0) + x' g_{l-1}(e_0)) &= Q(w' f_l(e_0) + x' g_l(e_0)) \\ &= Q(w' h_l(e_0) + x' k_l(e_0)) = \alpha. \end{aligned} \tag{5.4.41}$$

Here is our algorithm to find such vectors w', x'. First, find $z, t \in V$ such that $\langle V e_0,\ z,\ t \rangle = V$, and compute the scalars

$$\begin{aligned} \alpha_i &= (z,\ z f_i(e_0) + t g_i(e_0)) \quad (1 \le i \le 2l), \\ \beta_i &= (t,\ z h_i(e_0) + t k_i(e_0)) \quad (1 \le i \le 2l), \\ \gamma_i &= (z,\ z h_i(e_0) + t k_i(e_0)) \quad (0 \le i \le 2l), \\ \delta_1 &= Q(z), \\ \delta_2 &= Q(t). \end{aligned} \tag{5.4.42}$$

Now we aim to solve the following equations for $w', x' \in V$ and $b_i, c_i, d_i \in \mathbb{F}_q$:

$$\begin{aligned} z &= w' + \textstyle\sum_{i=0}^{l} b_i (w' h_{2i}(e_0) + x' k_{2i}(e_0)), \\ t &= \textstyle\sum_{i=0}^{l} c_i (w' f_{2i}(e_0) + x' g_{2i}(e_0)) + \sum_{i=0}^{2l} d_i (w' h_i(e_0) + x' k_i(e_0)) \end{aligned} \tag{5.4.43}$$

such that w', x' satisfy (5.4.40), (5.4.18) and (5.4.41). (Note the extra terms for $i = l$ in both equations, in comparison to (5.4.20).) The values $\alpha_i, \beta_i, \gamma_i, \delta_1, \delta_2$, combined with (5.4.43), (5.4.18) and (5.4.41), give $6l + 3$ quadratic equations in b_i, c_i, d_i.

To generate these equations, we follow the same procedure as in Case 4 of Sect. 5.4.1, with some small adjustments. We work in the same ring $R = F[W, X, E]/(E^{2l+1})$, where F is the field of rational functions over $\mathbb{F}_q$ in indeterminates b_i $(0 \le i \le l)$, c_i $(0 \le i \le l)$, and d_i $(0 \le i \le 2l)$. Define $V_i \in R$ as in (5.4.21), and for $r = \sum_{i=0}^{4l+1} r_i V_i$ and $s = \sum_{i=0}^{4l+1} s_i V_i \in R$, where each $r_i, s_i \in F$, define $IP(r, s)$ as in (5.4.22); also define

$$Quad(r) = \sum_{i=0}^{2l} r_i r_{4l+1-i} + \alpha(r_{l-1}^2 + r_l^2 + r_{3l+1}^2).$$

If we specialize down to $W \mapsto w$, $X \mapsto x$, $E \mapsto e$, then $V_0, \ldots, V_{4l+1}$ become the original basis vectors $w_{-2l}, \ldots, x_{2l}$, and, for $r_i \in \mathbb{F}_q$, we see that $r \in V$ and $Quad(r)$ is the value of the quadratic form $Q(r)$.

Now define the following elements of R:

$$
\begin{aligned}
Z &= W + \textstyle\sum_{i=0}^{l} b_i \left(W h_{2i}(E) + X k_{2i}(E) \right), \\
T &= \textstyle\sum_{i=0}^{l} c_i \left(W f_{2i}(E) + X g_{2i}(E) \right) + \sum_{i=0}^{2l} d_i \left(W h_i(E) + X k_i(E) \right).
\end{aligned}
$$

(As above, there are extra terms for $i = l$ in both elements, in comparison to Z, T as defined in Sect. 5.4.1.) Let $A_i, B_i, C_i \in F$ be as in (5.4.23), and define additional elements of F as follows:

$$Q_1(b, c, d) = Quad(Z), \;\; Q_2(b, c, d) = Quad(T).$$

It is easy to give explicit expressions for both.

$$Quad(Z) = \begin{cases} b_l & \text{for odd } l > 1 \\ b_l + \alpha b_{l/2}^2 & \text{for } l \text{ even} \\ b_1 + \alpha & \text{for } l = 1. \end{cases}$$

$$Quad(T) = \textstyle\sum_{i=0}^{l} c_i d_{2l-2i} + \alpha d_l^2 + \alpha c_{\lfloor l/2 \rfloor}^2.$$

The quadratic equations in b_i, c_i, d_i arising from (5.4.43) combined with (5.4.18), (5.4.41) and (5.4.42) are

$$
\begin{aligned}
A_i(b, c, d) &= \alpha_i \quad (1 \le i \le 2l), \\
B_i(b, c, d) &= \beta_i \quad (1 \le i \le 2l), \\
C_i(b, c, d) &= \gamma_i \quad (0 \le i \le 2l), \\
Q_1(b, c, d) &= \delta_1, \\
Q_2(b, c, d) &= \delta_2.
\end{aligned}
\tag{5.4.44}
$$

We use a Gröbner basis algorithm to find solutions for b_i, c_i, d_i of these equations. Given such a solution, define

$$
\begin{aligned}
f(e_0) &= 1 + \textstyle\sum_{i=0}^{l} b_i h_{2i}(e_0), \\
g(e_0) &= \textstyle\sum_{i=0}^{l} b_i k_{2i}(e_0), \\
r(e_0) &= \textstyle\sum_{i=0}^{l} c_i f_{2i}(e_0) + \sum_{i=0}^{2l} d_i h_i(e_0), \\
s(e_0) &= \textstyle\sum_{i=0}^{l} c_i g_{2i}(e_0) + \sum_{i=0}^{2l} d_i k_i(e_0).
\end{aligned}
$$

By (5.4.43),

$$z = w' f(e_0) + x' g(e_0), \;\; t = w' r(e_0) + x' s(e_0),$$

and we solve for w', x' exactly as in (5.4.26). This gives the required vectors w', x', completing this case $u = W_\alpha(2l + 1)$.

General Case
Suppose $g, h \in \mathrm{O}(V)$ are conjugate to a unipotent class representative u as in (5.1.1). Our algorithm mirrors that used in the general case of Sect. 5.4.1: we first replace summands of u of the form (5.4.28) by summands (5.4.29) to obtain a conjugate element u', and then compute $x \in \mathrm{O}(V)$ such that $g^x = h$.

If the class $u^{\mathrm{O}(V)}$ does not split in $\Omega(V)$, then g and h are $\Omega(V)$-conjugate, and we can obtain an element of $\Omega(V)$ conjugating g to h by adjusting x (if necessary) by an element of $\mathrm{O}(V) \setminus \Omega(V)$ that centralizes g. Finally, suppose the class $u^{\mathrm{O}(V)}$ splits in $\Omega(V)$ (so that $u = \bigoplus W(2k_i)^{[a_i]}$); if g and h are $\Omega(V)$-conjugate, then the conjugating element x automatically lies in $\Omega(V)$.

This completes our analysis of constructing conjugating elements for the orthogonal groups in characteristic 2.

5.5 Some Examples

We close this chapter by listing in Tables 5.3, 5.4, and 5.5 the unipotent class representatives and centralizer orders for 8-dimensional symplectic and orthogonal groups defined over fields of even size q. The structures of the centralizers are given by Theorem 5.2.1. Notation is as in earlier sections. Recall that α is a fixed element of $\mathbb{F}_q$ such that $x^2 + x + \alpha$ is irreducible in $\mathbb{F}_q[x]$.

Table 5.3 Unipotent class representatives in $G = \mathrm{Sp}_8(q)$, q even

Representative u	$\lvert C_G(u)\rvert$
$W(1)^{[4]}$	$\lvert G\rvert$
$W(4)$	$q^7\lvert\mathrm{Sp}_2(q)\rvert$
$W(3) \oplus W(1)$	$q^{10}\lvert\mathrm{Sp}_2(q)\rvert\,\lvert\mathrm{O}_2^+(q)\rvert$
$W_\alpha(3) \oplus W(1)$	$q^{10}\lvert\mathrm{Sp}_2(q)\rvert\,\lvert\mathrm{O}_2^-(q)\rvert$
$W(3) \oplus V(2)$	$q^9\lvert\mathrm{Sp}_2(q)\rvert$
$W(2)^{[2]}$	$q^{10}\lvert\mathrm{Sp}_4(q)\rvert$
$W(2) \oplus W(1) \oplus V(2)$	$q^{12}\lvert\mathrm{Sp}_2(q)\rvert^2$
$W(2) \oplus V(2)^{[2]}$	$q^{13}\lvert\mathrm{Sp}_2(q)\rvert$
$W(2) \oplus V_\beta(4)$ $(\beta \in \{1, \alpha\})$	$2q^9\lvert\mathrm{Sp}_2(q)\rvert$
$W(1)^{[3]} \oplus V(2)$	$q^7\lvert\mathrm{Sp}_6(q)\rvert$
$W(1)^{[2]} \oplus W(2)$	$q^{11}\lvert\mathrm{Sp}_2(q)\rvert\,\lvert\mathrm{Sp}_4(q)\rvert$
$W(1)^{[2]} \oplus V(2)^{[2]}$	$q^{12}\lvert\mathrm{Sp}_4(q)\rvert$
$W(1)^{[2]} \oplus V_\beta(4)$ $(\beta \in \{1, \alpha\})$	$2q^6\lvert\mathrm{Sp}_4(q)\rvert$
$W(1) \oplus V(4) \oplus V(2)$	$q^9\lvert\mathrm{Sp}_2(q)\rvert$
$W(1) \oplus V_\beta(6)$ $(\beta \in \{1, \alpha\})$	$2q^5\lvert\mathrm{Sp}_2(q)\rvert$
$V(4) \oplus V(2)^{[2]}$	q^{10}
$V(4) \oplus V_\beta(4)$ $(\beta \in \{1, \alpha\})$	$2q^8$
$V_\beta(6) \oplus V(2)$ $(\beta \in \{1, \alpha\})$	$2q^6$
$V_\beta(8)$ $(\beta \in \{1, \alpha\})$	$2q^4$

Table 5.4 Unipotent class representatives in $G = \Omega_8^+(q)$, q even

Representative u	$\lvert C_G(u)\rvert$
$W(1)^{[4]}$	$\lvert G\rvert$
$W(4),\ W(4)'$	$q^5\lvert \mathrm{Sp}_2(q)\rvert$
$W(2)^{[2]},\ W(2)\oplus W(2)'$	$q^6\lvert \mathrm{Sp}_4(q)\rvert$
$W(3)\oplus W(1)$	$2q^8(q-1)^2$
$W_\alpha(3)\oplus W_\alpha(1)$	$2q^8(q+1)^2$
$W(2)\oplus W(1)^{[2]}$	$q^9\lvert \mathrm{Sp}_2(q)\rvert\,\lvert \Omega_4^+(q)\rvert$
$W(2)\oplus V(2)^{[2]}$	$q^9\lvert \mathrm{Sp}_2(q)\rvert$
$W(1)^{[2]}\oplus V(2)^{[2]}$	$q^6\lvert \mathrm{Sp}_4(q)\rvert$
$W(1)\oplus V(4)\oplus V(2)$	$q^5\lvert \mathrm{Sp}_2(q)\rvert$
$V(4)^{[2]}$	q^6
$V(6)\oplus V(2)$	$2q^4$
$V_\alpha(6)\oplus V_\alpha(2)$	$2q^4$

Table 5.5 Unipotent class representatives in $G = \Omega_8^-(q)$, q even

Representative u	$\lvert C_G(u)\rvert$
$W(1)^{[4]}$	$\lvert G\rvert$
$W_\alpha(3)\oplus W(1)$	$2q^8(q^2-1)$
$W(3)\oplus W_\alpha(1)$	$2q^8(q^2-1)$
$W(2)\oplus W_\alpha(1)\oplus W(1)$	$q^9\lvert \mathrm{Sp}_2(q)\rvert\,\lvert \Omega_4^-(q)\rvert$
$W(2)\oplus V_\alpha(2)\oplus V(2)$	$q^9\lvert \mathrm{Sp}_2(q)\rvert$
$W(1)^{[2]}\oplus V_\alpha(2)\oplus V(2)$	$q^6\lvert \mathrm{Sp}_4(q)\rvert$
$W(1)\oplus V(4)\oplus V_\alpha(2)$	$q^5\lvert \mathrm{Sp}_2(q)\rvert$
$V_\alpha(4)\oplus V(4)$	q^6
$V_\alpha(6)\oplus V(2)$	$2q^4$
$V(6)\oplus V_\alpha(2)$	$2q^4$

Chapter 6
Semisimple Classes

Having handled unipotent conjugacy classes of the classical groups in Chaps. 4 and 5, we now consider semisimple classes. In the next chapter we combine the semisimple and unipotent analysis with the Jordan decomposition to deal with general conjugacy classes.

6.1 Semisimple Classes and Centralizers

Here we use the notation for classical groups introduced in Chap. 3. Let $F = \mathbb{F}_{q^u}$ with $u = 1$ or 2, and let $\lambda \mapsto \bar{\lambda}$ be the field automorphism of order u. Let V be an n-dimensional vector space over F and let β be a non-degenerate alternating, symmetric or hermitian form on V, and Q a non-degenerate quadratic form. Let $\mathcal{C}$ be the isometry group $\mathcal{C}(\beta)$ or $\mathcal{C}(Q)$.

Let $X \in \mathrm{GL}(V)$ be semisimple. Theorem 3.2.6 gives necessary and sufficient conditions for X to be similar to an element of $\mathcal{C}$—that is, for the intersection $X^{\mathrm{GL}(V)} \cap \mathcal{C}$ to be non-empty, where $X^{\mathrm{GL}(V)}$ denotes the conjugacy class of X in $\mathrm{GL}(V)$. Our next theorem determines precisely how this intersection splits into $\mathcal{C}$-classes.

Recall from Sect. 3.1 that for q odd, there are, up to congruence, two non-degenerate symmetric bilinear forms on a given vector space U over $\mathbb{F}_q$, and these are distinguished by their discriminants. We refer to these as the two *types* of such forms on U. If U has even dimension, then we also refer to the two types as plus and minus, in accordance with the description of orthogonal groups in Sect. 3.1.

Theorem 6.1.1 *Let $\mathcal{C} = \mathcal{C}(\beta)$ or $\mathcal{C}(Q)$ be an isometry group as above. Let $X_1, X_2 \in \mathcal{C}$ be semisimple.*

G. De Franceschi et al., *Conjugacy in Finite Classical Groups*, Springer Monographs in Mathematics, https://doi.org/10.1007/978-3-031-86461-2_6

(i) *If $\mathcal{C}$ is a symplectic group, a unitary group, or an orthogonal group in characteristic 2, then X_1 and X_2 are conjugate in $\mathcal{C}$ if and only if they are similar.*
(ii) *If $\mathcal{C}$ is an orthogonal group in odd characteristic, then X_1 and X_2 are conjugate in $\mathcal{C}$ if and only if they are similar and the forms induced by β on the eigenspaces for the eigenvalues $+1$ and -1 have the same type.*

Simultaneously, we prove the following result describing the centralizers of semisimple elements in $\mathcal{C}$. Recall, from Chap. 3, that for $X \in \mathcal{C}$, the generalized elementary divisors of X lie in the set $\Phi = \Phi_1 \cup \Phi_2 \cup \Phi_3$ of polynomials listed in Definition 3.2.4.

Theorem 6.1.2 *Let $\mathcal{C} = \mathcal{C}(\beta)$ or $\mathcal{C}(Q)$ as above, and let $X \in \mathcal{C}$ be semisimple. For every polynomial f in Φ, let m_f be the multiplicity of f as a generalized elementary divisor of X, let $d_f = u \deg(f)/2$ (where $u = 2$ if $\mathcal{C}$ is unitary and $u = 1$ otherwise), and let γ_f denote the restriction of the form β or Q to the generalized eigenspace $\ker(f(X))$. Then*

$$C_{\mathcal{C}}(X) \cong \prod_{f \in \Phi_1} \mathcal{C}(\gamma_f) \times \prod_{f \in \Phi_2} \mathrm{GL}_{m_f}(q^{d_f}) \times \prod_{f \in \Phi_3} \mathrm{GU}_{m_f}(q^{d_f}), \tag{6.1.1}$$

where the products run over all generalized elementary divisors of X.

Wall [33] proved Theorem 6.1.1 for symplectic groups in odd characteristic. We extend his approach to all sesquilinear and quadratic forms in arbitrary characteristic. In Sect. 6.1.1 we prove both theorems for sesquilinear forms, and complete their proof in Sect. 6.1.2 for orthogonal groups in characteristic 2.

6.1.1 Sesquilinear Forms

Let β be a non-degenerate sesquilinear form as above, and for $g \in \mathrm{GL}(V)$, define a form βg on V by

$$\beta g\,(u, v) = \beta(ug, vg) \quad \text{for all } u, v \in V.$$

Define $\mathcal{L} = \{\beta g : g \in \mathrm{GL}(V)\}$. For $X \in \mathcal{C}(\beta)$, let $\mathcal{L}^X = \{\gamma \in \mathcal{L} : \gamma X = \gamma\}$, the set of forms in $\mathcal{L}$ fixed by X. Let $X^{\mathrm{GL}(V)}$ be the conjugacy class of X in $\mathrm{GL}(V)$, and denote by $\mathcal{M}_X$ the set of $\mathcal{C}(\beta)$-conjugacy classes in $X^{\mathrm{GL}(V)} \cap \mathcal{C}(\beta)$. In the statement of the next lemma, ${}^gX$ denotes the conjugate gXg^{-1}.

Lemma 6.1.3 *Let $X \in \mathcal{C}(\beta)$. There is a bijection from $\mathcal{M}_X$ to the set of orbits of $C_{\mathrm{GL}(V)}(X)$ on $\mathcal{L}^X$. For ${}^gX \in \mathcal{C}(\beta)$ (where $g \in \mathrm{GL}(V)$), this bijection sends the conjugacy class $({}^gX)^{\mathcal{C}(\beta)}$ to the orbit of βg under the action of $C_{\mathrm{GL}(V)}(X)$. In particular, if $X_1, X_2 \in \mathcal{C}(\beta)$ are similar to X, with $X_i = {}^{g_i}X$ (for $g_i \in \mathrm{GL}(V)$,*

$i = 1, 2$), then X_1, X_2 are $\mathcal{C}(\beta)$-conjugate if and only if the forms βg_1, βg_2 are in the same $C_{\mathrm{GL}(V)}(X)$-orbit.

Proof The given map is a well-defined bijection from $\mathcal{M}_X$ to the set of orbits of $C_{\mathrm{GL}(V)}(X)$ on $\mathcal{L}^X$. □

We next identify some reductions we employ in the proof of Theorems 6.1.1 and 6.1.2. Let $X \in \mathcal{C}(\beta)$, and let $f_1, \ldots, f_h \in \Phi$ be the distinct generalized elementary divisors of X. Let the minimal polynomial of X be $f(t) = f_1(t)^{e_1} \cdots f_h(t)^{e_h}$, with generalized eigenspaces $V_i = \ker(f_i(X)^{e_i})$. Relative to a basis of V which is a union of bases of the V_i, the matrix of X is

$$\begin{pmatrix} X_1 & & & \\ & X_2 & & \\ & & \ddots & \\ & & & X_h \end{pmatrix},$$

where X_i is the matrix of the restriction of X to V_i. Moreover, the matrix of every form in $\mathcal{L}^X$ has shape

$$\begin{pmatrix} B_{11} & \cdots & B_{1h} \\ \vdots & \ddots & \vdots \\ B_{h1} & \cdots & B_{hh} \end{pmatrix},$$

where $X_i B_{ij} X_j^* = B_{ij}$, and $B_{ij} = \varepsilon B_{ji}^*$ for every $1 \le i, j \le h$ (where $\varepsilon = -1$ in the symplectic case and $\varepsilon = 1$ in the other cases). More generally, for every $g \in F[t]$,

$$g(X_i)B_{ij} = B_{ij}g(X_j^{*-1}). \tag{6.1.2}$$

Lemma 6.1.4 *With the above notation, if $f_i \neq f_j^*$, then $B_{ij} = 0$.*

Proof The polynomial $f_i^{e_i}$ is the minimal polynomial of X_i, so, taking $g = f_i$ in (6.1.2), we get $0 = f_i(X_i)^{e_i} B_{ij} = B_{ij} f_i(X_j^{*-1})^{e_i}$. Since f_i and f_j^* are coprime, $f_i(X_j^{*-1})^{e_i}$ is non-singular. Hence $B_{ij} f_i(X_j^{*-1})^{e_i} = 0$ implies $B_{ij} = 0$. □

This lemma implies that the matrix of every form in $\mathcal{L}^X$ has block diagonal shape

$$\begin{pmatrix} B_1 & & \\ & \ddots & \\ & & B_h \end{pmatrix}.$$

Thus the heart of the proof of Theorems 6.1.1 and 6.1.2 is the case where $h = 1$: namely, X has precisely one generalized elementary divisor. We handle this case in the next result. Since X is semisimple, $e_i = 1$ for all i.

Proposition 6.1.5 *Let $X \in \mathcal{C}(\beta)$ be semisimple, and assume that X has minimal polynomial $f \in \Phi$ which occurs with multiplicity m_f. Let $d_f = u\deg(f)/2$.*

(i) *If $f \in \Phi_1$, then X is a scalar and $C_{\mathcal{C}(\beta)}(X) = \mathcal{C}(\beta)$.*
(ii) *If $f \in \Phi_2$, then $X^{\mathrm{GL}(V)} \cap \mathcal{C}(\beta) = X^{\mathcal{C}(\beta)}$, a single $\mathcal{C}(\beta)$-conjugacy class, and*

$$C_{\mathcal{C}(\beta)}(X) = \mathrm{GL}_{m_f}(q^{d_f}).$$

(iii) *If $f \in \Phi_3$, then $X^{\mathrm{GL}(V)} \cap \mathcal{C}(\beta) = X^{\mathcal{C}(\beta)}$, a single $\mathcal{C}(\beta)$-conjugacy class, and*

$$C_{\mathcal{C}(\beta)}(X) = \mathrm{GU}_{m_f}(q^{d_f}).$$

Proof

(i) This is clear.
(ii) Suppose $f \in \Phi_2$, so that $f = gg^*$ where g is irreducible and $g \neq g^*$. Then $V = \ker g(X) \oplus \ker g^*(X)$. Let $\gamma \in \mathscr{L}^X$. Relative to a suitable basis of V, the matrices of X and γ are

$$X = \begin{pmatrix} X_1 & 0 \\ 0 & X_1^{*-1} \end{pmatrix}, \quad B = \begin{pmatrix} B_{11} & B_{12} \\ \varepsilon B_{12}^* & B_{22} \end{pmatrix}. \tag{6.1.3}$$

By Lemma 6.1.4, $B_{11} = B_{22} = 0$. The identity $XBX^* = B$ implies $X_1 B_{12} X_1^{-1} = B_{12}$, so X_1 commutes with B_{12}. If we take

$$Y = \begin{pmatrix} B_{12} & 0 \\ 0 & \mathbb{I} \end{pmatrix}, \quad J = \begin{pmatrix} 0 & \mathbb{I} \\ \varepsilon\mathbb{I} & 0 \end{pmatrix},$$

where $\mathbb{I}$ is the identity matrix of the same dimension as X_1, then Y commutes with X, and $YJY^* = B$. The matrix J does not depend on B, so given forms in $\mathscr{L}^X$ with matrices B, B' and Y, Y' constructed as above, $B = YJY^* = (YY'^{-1})B'(YY'^{-1})^*$ and $YY'^{-1} \in C_{\mathrm{GL}(V)}(X)$. This proves that all forms in $\mathscr{L}^X$ are in the same $C_{\mathrm{GL}(V)}(X)$-orbit, and so $X^{\mathrm{GL}(V)} \cap \mathcal{C}(\beta) = X^{\mathcal{C}(\beta)}$ by Lemma 6.1.3.

It remains to compute the centralizer of X in $\mathcal{C}(\beta)$. For this we may suppose that the form β has matrix J, and that X_1 is a block diagonal matrix

$$\begin{pmatrix} C & & \\ & \ddots & \\ & & C \end{pmatrix},$$

where C is the companion matrix of g. Every $Y \in C_{\mathrm{GL}(V)}(X)$ has the form

$$\begin{pmatrix} Y_1 & 0 \\ 0 & Y_2 \end{pmatrix},$$

where Y_1 commutes with X_1 and Y_2 with X_1^{*-1}, and the centralizer of X_1 is isomorphic to $\mathrm{GL}_{m_f}(q^{d_f})$ by Proposition 2.2.1. If also $Y \in C(\beta)$, then $YJY^* = J$, giving $Y_2 = (Y_1^*)^{-1}$. Hence Y_1 can be an arbitrary element of the centralizer of X_1 and it determines Y_2 uniquely. Thus $C_{C(\beta)}(X) = \mathrm{GL}_{m_f}(q^{d_f})$, completing the proof of (ii).

(iii) This case requires more effort. Suppose $f \in \Phi_3$, so that f is irreducible of degree $d \geq 2$ and $f = f^*$. For convenience write $h = m_f$, the multiplicity of the generalized elementary divisor f of X. Observe that X, relative to a suitable basis of V, has a block diagonal matrix

$$\begin{pmatrix} R & & \\ & \ddots & \\ & & R \end{pmatrix},$$

where R has minimal polynomial f and there are h diagonal blocks. By Proposition 2.2.1, $C_{\mathrm{GL}(V)}(X) \cong \mathrm{GL}_h(q^{2d_f})$, and Y commutes with X if and only if Y is non-singular and has the block matrix shape

$$\begin{pmatrix} & \vdots & \\ \cdots & f_{ij}(R) & \cdots \\ & \vdots & \end{pmatrix} \tag{6.1.4}$$

for some $f_{ij} \in F[t]$; in the unitary case, where $F = \mathbb{F}_{q^2}$, we can take $f_{ij} \in \mathbb{F}_q[t]$. Every form in $\mathscr{L}^X$ has block matrix

$$B = \begin{pmatrix} B_{11} & \cdots & B_{1h} \\ \vdots & \ddots & \vdots \\ B_{h1} & \cdots & B_{hh} \end{pmatrix},$$

and the equation $XBX^* = B$ is equivalent to

$$RB_{ij}R^* = B_{ij} \text{ for } 1 \leq i, j \leq h. \tag{6.1.5}$$

Since $f = f^*$, it follows that R^* is similar to R^{-1}, so there exists $T \in \mathrm{GL}_d(F)$ such that

$$R^* = T^{-1}R^{-1}T. \tag{6.1.6}$$

Thus (6.1.5) may be rewritten as

$$R(B_{ij}T^{-1}) = (B_{ij}T^{-1})R.$$

This shows that $B_{ij}T^{-1}$ belongs to the centralizer of R, and so $B_{ij} = f_{ij}(R)T$ for a certain polynomial $f_{ij} \in F[t]$; as above, in the unitary case we can take $f_{ij} \in \mathbb{F}_q[t]$. We obtain the equation

$$B = H\mathcal{T}, \tag{6.1.7}$$

where $H = (f_{ij}(R)) \in C_{\mathrm{GL}(V)}(X)$ and $\mathcal{T} = T \oplus \cdots \oplus T$.

Claim 1 The matrix T in (6.1.6) can be chosen such that $T = \varepsilon T^*$.

Proof of Claim 1 If $R^* = T^{-1}R^{-1}T$, then T can be replaced by $g(R)T$ for some $g \in F[t]$ such that $g(R)$ is invertible. Note that R has irreducible minimal polynomial $f(t)$, so every non-zero $g(R)$ is in fact invertible. So the aim is to prove that there exists $g(t)$ such that $g(R)T = \varepsilon(g(R)T)^*$ and $g(R) \neq 0$. From (6.1.6) we see that $R^* = T^{*-1}R^{-1}T^*$. Hence T^*T^{-1} commutes with R, so $T^* = \phi(R)T$ for some $\phi \in F[t]$.

If $T = -\varepsilon T^*$ and $RT = -\varepsilon(RT)^*$, then

$$RT = -\varepsilon(RT)^* = (-\varepsilon T^*)R^* = TR^* = R^{-1}T$$

by (6.1.6). But this implies $R = R^{-1}$ and so $R^2 = 1$, contradicting the assumption that $f \in \Phi_3$. Thus at least one of $T \neq -\varepsilon T^*$ and $RT \neq -\varepsilon(RT)^*$ holds.

If $T \neq -\varepsilon T^*$, then choose $g(t) = 1 + \varepsilon\phi(t)$ and deduce that

$$g(R)T = (1 + \varepsilon\phi(R))T = T + \varepsilon T^*.$$

Then $g(R)T = \varepsilon(g(R)T)^*$, as required.

If $RT \neq -\varepsilon(RT)^*$, then take $\psi \in F[t]$ such that $\psi(R) = R^{-1}$ and let $g(t) = t + \varepsilon\psi(t)\phi(t)$. Then

$$g(R)T = RT + \varepsilon R^{-1}T^* = RT + \varepsilon T^*R^* = RT + \varepsilon(RT)^*,$$

and again $g(R)T = \varepsilon(g(R)T)^*$. Claim 1 is now proved.

We observed in (6.1.4) that if $Y \in C_{\mathrm{GL}(V)}(X)$, then Y is a block matrix $(\phi_{ij}(R))$ (where $\phi_{ij} \in \mathbb{F}_q[t]$), so it can be identified with a matrix in $\mathrm{GL}_h(E)$, where E is the field $\mathbb{F}_q[t]/(f)$ in the symplectic and orthogonal cases, and E is the field $\mathbb{F}_q[t]/(f\bar{f})$ in the unitary case. The mapping $\phi_{ij}(R) \mapsto \phi_{ij}(R^{-1})$ is a field automorphism of E of order 2. For $Y = (\phi_{ij}(R)) \in \mathrm{GL}_h(E)$, define

$$Y^\dagger := (\phi_{ji}(R^{-1})). \tag{6.1.8}$$

The map $Y \mapsto Y^\dagger$ is a "conjugate-transpose" involutory automorphism of $\mathrm{GL}_h(E)$, where the transpose sends $(\phi_{ij}(R)) \mapsto (\phi_{ji}(R))$ and the conjugate is the automorphism $(\phi_{ij}(R)) \mapsto (\phi_{ij}(R^{-1}))$.

Claim 2 Let $B = H\mathcal{T}$ as in (6.1.7). Then $H = H^\dagger$. Moreover, if $Y \in C_{\mathrm{GL}(V)}(X) \cong \mathrm{GL}_h(E)$, then $YBY^* = YHY^\dagger\mathcal{T}$.

Proof of Claim 2 Since B is the matrix of a sesquilinear form in $\mathscr{L}$, we conclude that $B = \varepsilon B^*$. By (6.1.7), $B^* = \mathcal{T}^*H^*$, and by Claim 1, we can suppose that $\mathcal{T} = \varepsilon\mathcal{T}^*$. Using this and (6.1.6), we deduce that

$$\begin{aligned}\varepsilon B^* &= \varepsilon\mathcal{T}^*H^* \\ &= (\varepsilon T^* f_{ji}(R^*)) \\ &= (T f_{ji}(R^*)) \\ &= (f_{ji}(R^{-1})T) \\ &= H^\dagger\mathcal{T}.\end{aligned}$$

Since $B = \varepsilon B^*$, this implies that $H\mathcal{T} = H^\dagger\mathcal{T}$. Since $\mathcal{T}$ is invertible, $H = H^\dagger$.

Now consider the second assertion in Claim 2. Recall that $H = (f_{ij}(R)) \in C_{\mathrm{GL}(V)}(X)$. Write $Y = (\phi_{ij}(R))$ for $\phi_{ij} \in \mathbb{F}_q[t]$. Now

$$\begin{aligned}YBY^* = YH\mathcal{T}Y^* &= (\phi_{ij}(R))\,(f_{ij}(R)T)\,(\phi_{ij}(R))^* \\ &= \Big(\sum_{\lambda,\mu}\phi_{i\lambda}(R)f_{\lambda\mu}(R)T\phi_{j\mu}(R^*)\Big) \\ &= \Big(\sum_{\lambda,\mu}\phi_{i\lambda}(R)f_{\lambda\mu}(R)\phi_{j\mu}(R^{-1})T\Big) \\ &= YHY^\dagger\mathcal{T}.\end{aligned}$$

This completes the proof of Claim 2.

We now finish the proof of part (iii) of the proposition. If B_1 and B_2 are the matrices of forms in $\mathscr{L}^X$, with $B_1 = H_1\mathcal{T}$ and $B_2 = H_2\mathcal{T}$, then H_1 and H_2, viewed as matrices in $\mathrm{GL}_h(E)$, are hermitian, by Claim 2. Thus they are congruent: there exists $Y \in \mathrm{GL}_h(E) \cong C_{\mathrm{GL}(V)}(X)$ such that $H_1 = YH_2Y^\dagger$. By Claim 2, this implies

$$B_1 = H_1\mathcal{T} = YH_2Y^\dagger\mathcal{T} = YB_2Y^*.$$

There is only one orbit of forms in $\mathscr{L}^X$ under the action of $C_G(X)$, and hence $X^{\mathrm{GL}(V)} \cap \mathcal{C}(\beta) = X^{\mathcal{C}(\beta)}$ by Lemma 6.1.3.

For the final assertion, we must identify $C_{\mathcal{C}(\beta)}(X)$. Let B be the matrix of the form β, and let $B = H\mathcal{T}$ as in (6.1.7). By Claim 2, H is hermitian (with respect to the automorphism $\dagger$). Consider $W = E^h$ as an h-dimensional unitary space with respect to the form with matrix H. If $Y \in C_{\mathrm{GL}(V)}(X)$, then

$$\begin{aligned} Y \in \mathcal{C}(\beta) &\Leftrightarrow YBY^* = B \\ &\Leftrightarrow YHY^{\dagger}\mathcal{T} = H\mathcal{T} \\ &\Leftrightarrow YHY^{\dagger} = H. \end{aligned}$$

Hence the centralizer of X in $\mathcal{C}(\beta)$ is isomorphic to $\mathrm{GU}(W)$. This is the group $\mathrm{GU}_{m_f}(q^{d_f})$ in the conclusion of (iii). □

Proof of Theorems 6.1.1 and 6.1.2 for $\mathcal{C} = \mathcal{C}(\beta)$

Let $X \in \mathcal{C}(\beta)$ be semisimple, and let $f(t) = f_1(t) \cdots f_h(t)$ be the minimal polynomial of X, where each $f_i \in \Phi$ (and $f_1, \ldots, f_h$ are distinct). Let $V_i = \ker f_i(X)$. Now X, relative to a basis of V which is a union of bases of the V_i, has block diagonal matrix $\bigoplus_{i=1}^h X_i$, where X_i is the matrix of the restriction of X to V_i. Moreover, by Lemma 6.1.4, the matrix of β has the shape $B = \bigoplus_{i=1}^h B_i$, and every $Y \in C_{\mathrm{GL}(V)}(X)$ has the shape $\bigoplus_{i=1}^h Y_i$, where $[Y_i, X_i] = 1$ for all i. Also $Y \in \mathcal{C}(\beta)$ if and only if each Y_i fixes the form β_i with matrix B_i. Hence

$$C_{\mathcal{C}(\beta)}(X) = \prod_i C_{\mathcal{C}(\beta_i)}(X_i).$$

Each factor $C_{\mathcal{C}(\beta_i)}(X_i)$ is given by Proposition 6.1.5, and hence $C_{\mathcal{C}(\beta)}(X)$ is as in the conclusion of Theorem 6.1.2, completing the proof of that theorem.

We now prove Theorem 6.1.1. Let $X' \in \mathcal{C}(\beta)$ be similar to X, so $X' = X^g$ for some $g \in \mathrm{GL}(V)$. Then $X \in \mathcal{C}(\beta g)$, and, writing matrices with respect to the above basis, the matrix of βg is $B' = gBg^*$. By Lemma 6.1.3, X and X' are conjugate in $\mathcal{C}(\beta)$ if and only if β and βg are in the same $C_{\mathrm{GL}(V)}(X)$-orbit: namely, B and B' are congruent in $C_{\mathrm{GL}(V)}(X)$. As above, since X fixes βg, the matrix B' has block diagonal form $B' = \bigoplus_{i=1}^h B_i'$.

Assume now that, for all i, the forms B_i and B_i' have the same type: both are symplectic, or both are unitary, or both are orthogonal of the same discriminant. This assumption holds if β is symplectic or unitary. By Proposition 6.1.5, B_i and B_i' are congruent in $C_{\mathrm{GL}(V_i)}(X_i)$, so there exists $Y_i \in C_{\mathrm{GL}(V_i)}(X_i)$ such that $Y_i B_i Y_i^* = B_i'$. Then $Y = \bigoplus_{i=1}^h Y_i$ is in $C_{\mathrm{GL}(V)}(X)$ and satisfies $YBY^* = B'$. Hence X and X' are conjugate in $\mathcal{C}(\beta)$. This completes the proof of part (i) of Theorem 6.1.1 for sesquilinear forms.

It remains to consider the case where β is a symmetric form in odd characteristic. By Theorem 3.2.6, if $f_i \in \Phi_2 \cup \Phi_3$, then B_i and B_i' have the same type. If $f_i = t \pm 1 \in \Phi_1$, then B_i can be either of the two possible types. If at most one of $t + 1$ and $t - 1$ is an elementary divisor for X, then the types of B_i and B_i' are uniquely determined, and they coincide, so B and B' are congruent in $C_{\mathrm{GL}(V)}(X)$. However, if both $t + 1$ and $t - 1$ are elementary divisors for X, say $f_1 = t + 1$ and $f_2 = t - 1$, then B_1', B_2' can be either the same types as B_1, B_2, or the opposite types. Thus in this case the conjugacy class $X^{\mathrm{GL}(V)} \cap \mathcal{C}(\beta)$ splits into two $\mathcal{C}(\beta)$-classes.

This completes the proof of Theorems 6.1.1 and 6.1.2 for $\mathcal{C} = \mathcal{C}(\beta)$.

6.1.2 Orthogonal Groups in Characteristic 2

We now prove Theorems 6.1.1 and 6.1.2 for the orthogonal groups in characteristic 2. Let $\mathcal{C} = \mathcal{C}(Q) \cong \mathrm{O}_{2m}^{\epsilon}(q)$, where Q is a non-degenerate quadratic form on a $2m$-dimensional vector space V over $F = \mathbb{F}_q$, with $q = 2^k$. Let β_Q be the associated symplectic form on V. As in Sect. 6.1.1, the heart of the proof is the case where the elements in question (X_1, X_2 in Theorem 6.1.1, and X in Theorem 6.1.2) have precisely one elementary divisor. The next result is the analogue of Proposition 6.1.5.

Proposition 6.1.6 *Let $\mathcal{C}(Q) \cong \mathrm{O}_{2m}^{\epsilon}(q)$, and let β_Q be the associated symplectic form. Let $X \in \mathcal{C}(Q)$ be semisimple, and assume that X has minimal polynomial $f \in \Phi$.*

(i) *If $f \in \Phi_1$, then $X = I$ and $C_{\mathcal{C}(Q)}(X) = \mathcal{C}(Q)$.*
(ii) *If $f \in \Phi_2 \cup \Phi_3$, then $X^{\mathrm{GL}(V)} \cap \mathcal{C}(Q) = X^{\mathcal{C}(Q)}$, a single $\mathcal{C}(Q)$-conjugacy class, and*

$$C_{\mathcal{C}(Q)}(X) = C_{\mathcal{C}(\beta_Q)}(X).$$

Hence $C_{\mathrm{O}_{2m}^{\epsilon}(q)}(X) = C_{\mathrm{Sp}_{2m}(q)}(X)$ (which is given by Proposition 6.1.5).

Proof Part (i) is trivial. So assume $f \in \Phi_2 \cup \Phi_3$. Recall that Q can be represented by a matrix $A = (a_{ij})$ such that $Q(v) = vAv^{\mathrm{t}}$ for all $v \in V$.

Let $\mathrm{O}_{2m+1}(q)$ be the group of isometries for the quadratic form with matrix

$$\widehat{A} = \begin{pmatrix} 1 & 0 \\ 0 & A \end{pmatrix}.$$

Every matrix in $\mathrm{O}_{2m+1}(q)$ has shape

$$\widehat{Y} = \begin{pmatrix} 1 & 0 \\ v & Y \end{pmatrix}, \tag{6.1.9}$$

where v is a $2m$-dimensional column vector. Note that if $v = 0$, then Y is an isometry in $\mathcal{C}(A) = \mathrm{O}_{2m}^{\epsilon}(q)$. It is well known (see [15, 14.1]) that the map $\widehat{Y} \mapsto Y$ is an isomorphism between $\mathrm{O}_{2m+1}(q)$ and $\mathcal{C}(A + A^{\mathrm{t}}) \cong \mathrm{Sp}_{2m}(q)$, and the inverse of this isomorphism maps Y to a matrix of the above form, where $v = 0$ if and only if $Y \in \mathrm{O}_{2m}^{\epsilon}(q)$. We now use this isomorphism to describe the centralizer of X in

$\mathrm{O}_{2m}^{\epsilon}(q)$. The centralizer of X in $\mathrm{Sp}_{2m}(q)$ is isomorphic to the centralizer of $\widehat{X}$ in $\mathrm{O}_{2m+1}(q)$, where

$$\widehat{X} = \begin{pmatrix} 1 & 0 \\ 0 & X \end{pmatrix}.$$

Since $t + 1$ is not an elementary divisor of X, every element of the centralizer of $\widehat{X}$ in $\mathrm{O}_{2m+1}(q)$ has the form $\widehat{Y}$ as in (6.1.9) with $v = 0$ for some Y, and by the above observations, $Y \in \mathrm{O}_{2m}^{\epsilon}(q)$. Conversely, it is clear that for every $Y \in C_{\mathrm{O}_{2m}^{\epsilon}(q)}(X)$, the corresponding $\widehat{Y}$ is in the centralizer of $\widehat{X}$ in $\mathrm{O}_{2m+1}(q)$. This proves that $|C_{\mathrm{O}_{2m}^{\epsilon}(q)}(X)| = |C_{\mathrm{Sp}_{2m}(q)}(X)|$, so these centralizers are equal, since the first is obviously contained in the second.

Finally, if $X_1, X_2 \in \mathrm{O}_{2m}^{\epsilon}(q)$ are similar, both having minimal polynomial $f \in \Phi_2 \cup \Phi_3$, then they are conjugate in $\mathrm{Sp}_{2m}(q)$ by Proposition 6.1.5, and so $\widehat{X}_1$ and $\widehat{X}_2$ are conjugate in $\mathrm{O}_{2m+1}(q)$. Hence there exists

$$\widehat{Z} = \begin{pmatrix} 1 & 0 \\ v_Z & Z \end{pmatrix} \in \mathrm{O}_{2m+1}(q)$$

such that $\widehat{Z}^{-1}\widehat{X}_1\widehat{Z} = \widehat{X}_2$. Since $t+1$ is not an elementary divisor of X_1, we deduce that $\widehat{Z}$ preserves the diagonal block structure of $\widehat{X}_1$ and $\widehat{X}_2$, so $v_Z = 0$. This shows that $Z \in \mathrm{O}_{2m}^{\epsilon}(q)$, and so X_1 and X_2 are conjugate in $\mathrm{O}_{2m}^{\epsilon}(q)$. □

As in Sect. 6.1.1, the conclusions of Theorems 6.1.1 and 6.1.2 now follow.

6.1.3 Special and Omega Groups

We now determine the splitting of those semisimple classes in the isometry groups $\mathcal{C} = \mathrm{GU}_n(q)$ or $\mathrm{O}_n^{\epsilon}(q)$ that lie in the special groups $\mathscr{S} = \mathrm{SU}_n(q)$ or $\mathrm{SO}_n^{\epsilon}(q)$, or in the Omega group $\Omega = \Omega_n^{\epsilon}(q)$. This is a matter of computing the indices $|C_{\mathcal{C}}(X) : C_{\mathscr{S}}(X)|$ and $|C_{\mathcal{C}}(X) : C_{\Omega}(X)|$ for semisimple elements X of $\mathscr{S}$ and Ω. At the end of the section we show how to determine whether a given semisimple element of $\mathscr{S}$ lies in Ω.

Proposition 6.1.7

(i) *Let* $\mathcal{C} = \mathrm{GU}_n(q)$ *and* $\mathscr{S} = \mathrm{SU}_n(q)$. *If* $X \in \mathscr{S}$ *is semisimple, then* $|C_{\mathcal{C}}(X) : C_{\mathscr{S}}(X)| = q + 1$ *and* $X^{\mathcal{C}} = X^{\mathscr{S}}$.
(ii) *Let* $\mathcal{C} = \mathrm{O}_n^{\epsilon}(q)$ *and* $\mathscr{S} = \mathrm{SO}_n^{\epsilon}(q)$ *with* q *odd. If* $X \in \mathscr{S}$ *is semisimple, then* $|C_{\mathcal{C}}(X) : C_{\mathscr{S}}(X)| = c_S$, *where*

$$c_S = \begin{cases} 2, & \text{if } X \text{ has an eigenvalue } \pm 1, \\ 1, & \text{otherwise.} \end{cases}$$

The class $X^{\mathcal{C}}$ splits into $2/c_s$ conjugacy classes in $\mathcal{S}$ of equal size; if $c_S = 1$, then representatives are X and X^s for $s \in \mathcal{C} \setminus \mathcal{S}$. If n is odd, then $c_S = 2$ always.

(iii) *Let $\mathcal{S} = \mathrm{SO}_n^{\epsilon}(q)$ and $\Omega = \Omega_n^{\epsilon}(q)$. If $X \in \Omega$ is semisimple, then $|C_{\mathcal{S}}(X) : C_{\Omega}(X)| = c_{\Omega}$, where*

$$c_{\Omega} = \begin{cases} 2, & \text{if } X \text{ has an eigenvalue } \pm 1, \text{ or if } q \text{ is odd,} \\ 1, & \text{otherwise.} \end{cases}$$

The class $X^{\mathcal{S}}$ splits into $2/c_{\Omega}$ conjugacy classes in Ω of equal size; if $c_{\Omega} = 1$, then representatives are X and X^v for $v \in \mathcal{S} \setminus \Omega$. If n is odd, then $c_{\Omega} = 2$ always.

Proof

(i) We show that the centralizer $C_{\mathcal{C}}(X)$, given by (6.1.1), contains elements of all determinants in $D := \{\lambda \in \mathbb{F}_{q^2} : \lambda^{q+1} = 1\}$. This is clearly the case if X has a generalized elementary divisor $f \in \Phi_1$, since $C_{\mathcal{C}}(X)$ has a factor $\mathcal{C}(\gamma_f) = \mathrm{GU}_{m_f}(q)$. Consider a generalized elementary divisor $f = gg^* \in \Phi_2$, where g is irreducible and $g \neq g^*$. The corresponding factor of $C_{\mathcal{C}}(X)$ is $\mathrm{GL}_m(q^{2d})$, where $m = m_f$ and $d = \deg g$, and from the proof of Proposition 6.1.5 the factor consists of block diagonal matrices $(Y, (Y^*)^{-1})$ for $Y \in \mathrm{GL}_m(q^{2d})$. We have an embedding

$$\mathrm{GL}_m(q^{2d}) \leq \mathrm{GL}_{md}(q^2) < \mathrm{GU}_{2md}(q) \leq \mathcal{C},$$

and, via this embedding, $Y \in \mathrm{GL}_m(q^{2d})$ can have any determinant $\lambda \in \mathbb{F}_{q^2}^*$. Hence $(Y, (Y^*)^{-1})$ can have any determinant $\lambda\bar{\lambda}^{-1}$ in D, proving the result for generalized elementary divisors $f = gg^* \in \Phi_2$. Finally, consider a generalized elementary divisor $f \in \Phi_3$ of X, with corresponding factor $\mathrm{GU}_m(q^d)$ of $C_{\mathcal{C}}(X)$, where $d = \deg f$. Here d is odd by Proposition 3.2.5, and we have an embedding $\mathrm{GU}_m(q^d) \leq \mathrm{GU}_{md}(q)$, for which [21, (4.3.13)] shows that $\mathrm{GU}_m(q^d)$ contains elements of all determinants in D. This proves (i).

(ii) Note that $|\mathcal{C} : \mathcal{S}| = 2$, and the elements in the coset $\mathcal{C} \setminus \mathcal{S}$ are those of determinant -1. As in the proof of (i), we need to show that $C_{\mathcal{C}}(X)$ contains elements of determinant -1 if and only if X has an eigenvalue ± 1. The centralizer $C_{\mathcal{C}}(X)$ is given by (6.1.1). Let f be a generalized elementary divisor of X. If $f \in \Phi_1$, then X has an eigenvalue ± 1 and $C_{\mathcal{C}}(X)$ has a factor $\mathrm{O}_{m_f}(q)$, which has elements of determinant ± 1. On the other hand, if $f \in \Phi_2 \cup \Phi_3$, then the corresponding factors $\mathrm{GL}_m(q^d)$ and $\mathrm{GU}_m(q^d)$ (where $m = m_f, d = d_f$) lie in the special orthogonal group $\mathcal{S}$ via the embeddings

$$\mathrm{GL}_m(q^d) \leq \mathrm{SO}_{2m}^+(q^d) < \mathrm{SO}_{2md}^+(q)$$

and

$$\mathrm{GU}_m(q^d) \leq \mathrm{SO}^{\epsilon}_{2m}(q^d) < \mathrm{SO}^{\epsilon}_{2md}(q),$$

where $\epsilon = (-)^m$ (see [21, Props. 4.2.7, 4.3.18]). This proves (ii).

(iii) This follows as in (ii) once we establish that $\mathrm{GL}_m(q^d) \leq \Omega^+_{2md}(q)$ and $\mathrm{GU}_m(q^d) \leq \Omega^{\epsilon}_{2md}(q)$ if and only if q is even, where $m = m_f, d = d_f$ and $f \in \Phi_2 \cup \Phi_3$ as before. For even q these containments are clear, since neither $\mathrm{GL}_m(q^d)$ nor $\mathrm{GU}_m(q^d)$ has a subgroup of index 2.

Assume q is odd. Consider the embedding

$$\mathrm{GL}_m(q^d) \leq \mathrm{GL}_{md}(q) < \mathrm{SO}^+_{2md}(q) = \mathrm{SO}(V),$$

where $\mathrm{GL}_{md}(q)$ stabilizes two maximal totally isotropic subspaces $W = \langle e_1, \dots, e_{md}\rangle$ and $W' = \langle f_1, \dots, f_{md}\rangle$ (and e_i, f_i are in a standard hyperbolic basis). Consider $g \in \mathrm{GL}_{md}(q)$ sending $e_1 \mapsto \lambda e_1$, $f_1 \mapsto \lambda^{-1} f_1$ and fixing all other e_i, f_i, where $\lambda \in \mathbb{F}_q$ is a non-square. Now g can be expressed as the product $r_{e_1 - \lambda f_1} r_{e_1 - f_1}$ of the reflections in the vectors $e_1 - \lambda f_1$ and $e_1 - f_1$, and so has non-identity spinor norm. It follows that $g \notin \Omega(V)$, and hence no element in the subgroup $\mathrm{GL}_{md}(q)$ of determinant λ is in $\Omega(V)$, proving that $\mathrm{GL}_m(q^d) \not\leq \Omega^+_{2md}(q)$, as required. Now consider the embedding

$$\mathrm{GU}_m(q^d) \leq \mathrm{SO}^{\epsilon}_{2m}(q^d) < \mathrm{SO}^{\epsilon}_{2md}(q) = \mathrm{SO}(V)$$

where $d = d_f$ and $f \in \Phi_3$. Choose a natural subgroup $H = \mathrm{GU}_1(q^d)$ of the left-hand group, so $H \leq \mathrm{SO}^-_2(q^d) < \mathrm{SO}^-_{2d}(q) = \mathrm{SO}(W)$, fixing $W^\perp$ pointwise. From [21, Lemma 4.3.15] we see that $N_{\Omega(W)}(H)$ has order $d(q^d + 1)/2$, whereas $N_{\mathrm{SO}(W)}(H) = \mathrm{GU}_1(q^d).d$ has order $d(q^d + 1)$. Hence $H \not\leq \Omega(W)$, and it follows that $\mathrm{GU}_m(q^d) \not\leq \Omega^{\epsilon}_{2md}(q)$, as required. □

Finally, we record a consequence of the above proof that determines which semisimple elements of $\mathrm{SO}(V)$ are in $\Omega(V)$.

Lemma 6.1.8 *Let $V = V_n(q)$ be an orthogonal space with quadratic form Q, let $\mathcal{S} = \mathrm{SO}(V)$, and $\Omega = \Omega(V)$.*

(i) *Every element of odd order in $\mathcal{S}$ is also in Ω.*

(ii) *Let q be odd and let $x \in \mathcal{S}$ be semisimple. Assume that x has a unique generalized elementary divisor $f \in \Phi$ of degree d and f occurs with multiplicity m.*

  (a) *If $f = t - 1$, then $x = I \in \Omega$.*

  (b) *If $f = t + 1$, then $x = -I \in \Omega$ if and only if Q has square discriminant.*

(c) *Let* $f \in \Phi_2$ *with* $f = gg^*$, *where* $g \neq g^*$ *is irreducible, and let* C *be the companion matrix of* g. *Then* $x \in \Omega$ *if and only if either* m *is even, or* $\det(C)$ *is a square in* $\mathbb{F}_q^*$.

(d) *Let* $f \in \Phi_3$ *have degree* d *and companion matrix* C. *Then* $x \in \Omega$ *if and only if either* m *is even, or the order of* C *divides* $(q^{d/2}+1)/2$.

Proof Part (i) is obvious, as $|\mathscr{S} : \Omega| = 2$. Part (ii)(b) follows from [21, Prop. 2.5.13]. Finally, (ii)(c) and (ii)(d) follow from the argument for part (iii) of the previous proof. □

If $x \in \mathscr{S}$ is semisimple and has generalized elementary divisors $f_1, \ldots, f_k$, then we can decide its membership in Ω by applying the lemma to each f_i.

6.2 Representatives for Semisimple Classes

We show how to write down an explicit set of representatives for the semisimple conjugacy classes of classical groups. We give the matrices of both the representatives and the corresponding sesquilinear or quadratic forms with respect to an appropriate basis. We thank Donald Taylor for his assistance with this task.

Let β or Q be a sesquilinear or quadratic form on V, and let $\mathcal{C} = \mathcal{C}(\beta)$ or $\mathcal{C}(Q)$. By Lemma 6.1.4, every semisimple class in $\mathcal{C}$ has a block diagonal representative $X = \bigoplus_{i=1}^h X_i$, where each X_i has a single generalized elementary divisor f_i, and the form β or β_Q has corresponding matrix $B = \bigoplus_{i=1}^h B_i$. By Theorem 6.1.1, it is sufficient to write down, for each $f \in \Phi$, a matrix X_f, with generalized elementary divisor f of multiplicity 1 when $f \in \Phi_2 \cup \Phi_3$, and a form $\gamma_f = \beta_f$ or Q_f with matrix B_f such that $X_f \in \mathcal{C}(\gamma_f)$. We do this separately for the cases where $f \in \Phi_1$, Φ_2 or Φ_3.

Let $C(h)$ denote the companion matrix of $h(x)$, a monic polynomial of degree d.

Case $f \in \Phi_1$

Here X_f is a scalar matrix $\pm I$ in the symplectic and orthogonal cases, and is λI with $\lambda^{q+1} = 1$ in the unitary case, and B_f can be taken with respect to any basis (but we must allow both types for Q_f in the orthogonal case).

Case $f \in \Phi_2$

Let $f = gg^*$, where $g \neq g^*$ and g is irreducible. With respect to a suitable basis we take

$$X_f = \begin{pmatrix} C(g) & \\ & C(g)^{*-1} \end{pmatrix}, \quad B_f = \begin{pmatrix} \mathbb{O} & \mathbb{I} \\ \varepsilon\mathbb{I} & \mathbb{O} \end{pmatrix},$$

where ε is -1 (for $\gamma_f = \beta_f$ symplectic), 1 (for $\gamma_f = \beta_f$ symmetric or unitary), or 0 (for $\gamma_f = Q_f$ quadratic).

Case $f \in \Phi_3$

Symplectic case. Here $\deg f$ is even by Proposition 3.2.5, and f has constant term 1. Since $f = f^*$, we can write

$$f(t) = 1 + a_1 t + a_2 t^2 + \cdots + a_d t^d + a_{d-1} t^{d+1} + \cdots + a_1 t^{2d-1} + t^{2d}. \quad (6.2.1)$$

Define

$$X_f = C(f)^{\mathrm{t}}, \quad B_f = \begin{pmatrix} \mathbb{O} & -P^{\mathrm{t}} \\ P & \mathbb{O} \end{pmatrix},$$

where $C(f)$ is the companion matrix of f, and P is the $d \times d$ upper triangular matrix with constant upper diagonals:

$$P = \begin{pmatrix} 1 & a_1 & a_2 & \cdots & a_{d-2} & a_{d-1} \\ & 1 & a_1 & \ddots & \ddots & a_{d-2} \\ & & \ddots & \ddots & \ddots & \vdots \\ & & & \ddots & \ddots & a_2 \\ & & & & 1 & a_1 \\ & & & & & 1 \end{pmatrix}.$$

A computation shows that $X_f \in \mathcal{C}(B_f)$.

Quadratic Case. The degree of $f \in \Phi_3$ is even. If $\deg f = 2$, so $f = t^2 + at + 1$, then we take $X_f = C(f)$, and Q_f to be the quadratic form with matrix

$$B_f = \begin{pmatrix} 1 & -a \\ 0 & 1 \end{pmatrix}.$$

Then $X_f \in \mathcal{C}(Q_f)$.

Now suppose $\deg f > 2$, where f is defined in (6.2.1), and $d \geq 2$. Let $X_f = C(f)$, and define Q_f to be the quadratic form with $2d \times 2d$ matrix

$$B_f = \begin{pmatrix} 0 & A \\ 0 & 0 \end{pmatrix},$$

where the 0's are zero matrices of the appropriate sizes, A is the $(d+1)\times(d+1)$ matrix

$$\begin{pmatrix} 1 & b_0 & b_1 & \cdots & b_{d-1} \\ & 1 & \ddots & \ddots & \vdots \\ & & \ddots & \ddots & b_1 \\ & & & \ddots & b_0 \\ & & & & 1 \end{pmatrix},$$

and the coefficients b_i are defined as follows. If p is odd, then the vector of the b_i's satisfies the linear system

$$\begin{pmatrix} b_0 & \cdots & b_{d-1} \end{pmatrix} \begin{pmatrix} 2 & a_1 & a_2 & \cdots & a_{d-1} \\ & 1 & a_1 & \ddots & \vdots \\ & & \ddots & \ddots & a_2 \\ & & & \ddots & a_1 \\ & & & & 1 \end{pmatrix} = \begin{pmatrix} 2a_1 & a_2 - 1 & a_3 & a_4 & \cdots & a_d \end{pmatrix}.$$

If $p = 2$, then the vector of the b_i's satisfies the linear system

$$\begin{pmatrix} b_0 & \cdots & b_{d-1} \end{pmatrix} \begin{pmatrix} c_1 & a_1 & a_2 & \cdots & a_{d-1} \\ c_2 & 1 & a_1 & \ddots & \vdots \\ \vdots & & \ddots & \ddots & a_2 \\ \vdots & & & \ddots & a_1 \\ c_d & & & & 1 \end{pmatrix} = \begin{pmatrix} \delta & a_2 - 1 & a_3 & a_4 & \cdots & a_d \end{pmatrix},$$

where

$$c_i = \sum_{j=1}^{d+1-i} a_{j-1} a_{d+1-j} \qquad \text{for } i = 1, \ldots, d;$$

$$\delta = \sum_{j=0}^{d} a_j a_{d+j-1}$$

and we set $a_0 = 1$ and $a_{d+j} = a_{d-j}$ for $j = 1, \ldots, d$. A computation shows that $X_f \in C(Q_f)$.

Unitary Case. Here $\deg f$ is odd by Proposition 3.2.5. Since $f = f^*$ we can write

$$f(t) = a_0 + a_1 t + \cdots + a_d t^d + a_0 t^{d+1}(\overline{a}_d + \overline{a}_{d-1} t + \cdots + \overline{a}_0 t^d)$$

(an irreducible polynomial in $\mathbb{F}_{q^2}[t]$ with $a_0\overline{a}_0 = 1$). Choose $b_0 \in \mathbb{F}_{q^2}$ such that $b_0^{q-1} = a_0^{-1}$, and define

$$b_i := \overline{b}_0 \sum_{j=0}^{i} a_j \quad \text{for } 1 \leq i \leq d.$$

Choose also $c \in \mathbb{F}_{q^2}$ such that $c^{q+1} = b_d + \overline{b}_d$. Now define the $(2d+1) \times (2d+1)$ matrix

$$X_f = \left(\begin{array}{cccc|c|cccc}
1 & & & & & & & & \\
 & \ddots & & & & & & & \\
 & & & 1 & & & & & 1/\overline{b}_0 \\ \hline
 & & & & 1 & & & & -\overline{c}/\overline{b}_0 \\ \hline
b_0 & b_1 & \cdots & b_{d-1} & c & & & & -\overline{b}_d/\overline{b}_0 \\
 & & & & & 1 & & & -\overline{b}_{d-1}/\overline{b}_0 \\
 & & & & & & \ddots & & \vdots \\
 & & & & & & & 1 & -\overline{b}_1/\overline{b}_0
\end{array}\right).$$

Observe that X_f has characteristic and minimal polynomial f and preserves the hermitian form with matrix

$$B_f = \begin{pmatrix} & & 1 \\ & \cdot^{\cdot^{\cdot}} & \\ 1 & & \end{pmatrix}.$$

In Sect. 7.1, we need to compute the matrix of the form preserved by the companion matrix $C(f)$. We can compute this by a change of basis: if X_f and B_f are the matrices described above, and $P \in \mathrm{GL}(V)$ satisfies $PX_fP^{-1} = C(f)$, then $C(f)$ preserves the form PB_fP^*.

This completes the description of semisimple class representatives in the isometry groups $\mathcal{C} = \mathcal{C}(\beta)$ or $\mathcal{C}(Q)$. Representatives in $\mathscr{S}$ and Ω can be written down using Proposition 6.1.7.

6.3 Generators for the Centralizer of a Semisimple Element

Having identified the structure of the centralizer of a semisimple element in a classical isometry group in Theorem 6.1.2, we now show how to write down a generating set for the centralizer.

We assume that the following algorithms are available.

- (*Algorithm* 1) Given similar matrices $X, Y \in \mathrm{GL}_n(q)$, we determine explicitly $Z \in \mathrm{GL}_n(q)$ such that $Z^{-1}XZ = Y$. This follows from the Jordan basis algorithm: if J is the Jordan form of X and Y and $J = P_X X P_X^{-1} = P_Y Y P_Y^{-1}$ for $P_X, P_Y \in \mathrm{GL}(V)$, then $Z = P_X^{-1} P_Y$.
- (*Algorithm* 2) Given matrices B_1, B_2 of two non-degenerate sesquilinear or quadratic forms on $V = V_n(q^u)$ of the same type (where $u = 2$ for unitary forms, $u = 1$ otherwise), we determine explicitly $T \in \mathrm{GL}_n(q^u)$ such that $TB_1T^* = B_2$ (or $TB_1T^* - B_2$ is alternating in the case of quadratic forms).

Descriptions of such algorithms appear in [29] and [34] respectively. Algorithm 2 allows us to write a generating set for a classical group in any basis. Suppose a sesquilinear or quadratic form β has matrix B in a certain basis, and let B_0 be the matrix of β in the basis with respect to which the standard generators for $\mathcal{C}(\beta)$ are defined. If $\{Y_1, \dots, Y_r\}$ generates $\mathcal{C}(B_0)$, then $\{TY_1T^{-1}, \dots, TY_rT^{-1}\}$ generates $\mathcal{C}(B)$, where T is the matrix returned by Algorithm 2 such that $B = TB_0T^*$.

Let B be the matrix of a non-degenerate sesquilinear form on V. Let $X \in \mathcal{C}(B)$ be semisimple and let $f_1, \dots, f_h \in \Phi$ be the generalized elementary divisors of X. With respect to a suitable basis,

$$X = \begin{pmatrix} X_1 & & \\ & \ddots & \\ & & X_h \end{pmatrix}, \tag{6.3.1}$$

where each X_i is the matrix of the restriction of X to the generalized eigenspace $\ker(f_i(X))$. By Lemma 6.1.4, with respect to this basis,

$$B = \begin{pmatrix} B_1 & & \\ & \ddots & \\ & & B_h \end{pmatrix} \tag{6.3.2}$$

and each X_i is an isometry for B_i. If $Y \in C_{\mathcal{C}(B)}(X)$, then Y is the block diagonal sum of Y_i for $Y_i \in C_{\mathcal{C}(B_i)}(X_i)$, so we can write down generators for $C_{\mathcal{C}(B)}(X)$ once we have done this when X has a single generalized elementary divisor.

Suppose that X has a single generalized elementary divisor f, and let m be the multiplicity of f and $d = \deg f$.

Case 1: $f \in \Phi_1$. Here X is a scalar matrix, so its centralizer is $\mathcal{C}(B)$. A generating set for $\mathcal{C}(B)$ is referenced in Sect. 3.1.

Case 2: $f \in \Phi_2$. Here $f = gg^*$, where g is irreducible and $g \neq g^*$. Let $d' = d/2$ be the degree of g. As in Sect. 6.2, we choose a basis with respect to which

$$X = \begin{pmatrix} \widehat{X} & \mathbb{O} \\ \mathbb{O} & \widehat{X}^{*-1} \end{pmatrix},$$

where $\widehat{X}$ is a block diagonal sum of m copies of the companion matrix $C(g)$. By Lemma 6.1.4, the form preserved by X is

$$B = \begin{pmatrix} \mathbb{O} & A \\ \varepsilon A^* & \mathbb{O} \end{pmatrix}, \tag{6.3.3}$$

where $\varepsilon = -1$ in the symplectic case and 1 otherwise; and the centralizer of X in $\mathcal{C}(B)$ is the group of matrices of the form

$$\begin{pmatrix} Y & \mathbb{O} \\ \mathbb{O} & A^* Y^{*-1} A^{*-1} \end{pmatrix},$$

where $Y \in C_{\mathrm{GL}_{d'm}(q)}(\widehat{X}) \cong \mathrm{GL}_m(q^{d'})$. So, if Y_1, Y_2 are standard generators for $C_{\mathrm{GL}_{d'm}(q)}(\widehat{X})$, then the centralizer of X in $\mathcal{C}(B)$ is generated by

$$\begin{pmatrix} Y_i & \\ & A^* Y_i^{*-1} A^{*-1} \end{pmatrix} \quad (i = 1, 2).$$

Case 3: $f \in \Phi_3$. Here $f = f^*$ is irreducible. Let E be the field $\mathbb{F}_q[t]/(f)$ in the symplectic and orthogonal cases, and let $E = \mathbb{F}_q[t]/(f\bar{f})$ in the unitary case. We follow the argument in the proof of part (iii) of Proposition 6.1.5. Let R be the companion matrix of f and let $\varepsilon = -1$ if B is alternating, $\varepsilon = 1$ otherwise. Let X be the block diagonal sum of m copies of R. Using Algorithm 1 we find T such that $R^* = T^{-1}R^{-1}T$, and by Claim 1 in the proof of Proposition 6.1.5(iii), we can choose T such that $T = \varepsilon T^*$. Let $\mathcal{T}$ be the block diagonal sum of m copies of T. The matrix $H = B\mathcal{T}^{-1}$ lies in the centralizer of X, so it is the embedding into $\mathrm{GL}_{md}(F)$ of $\widetilde{H} \in \mathrm{GL}_m(E)$. By Claim 2 in the proof of Proposition 6.1.5(iii), $\widetilde{H}$ is hermitian and, if $Y \in C_{\mathcal{C}(B)}(X)$, then Y is the embedding into $\mathrm{GL}_{md}(F)$ of some $\widetilde{Y} \in \mathcal{C}(\widetilde{H})$. So, if $\widetilde{Y}_1, \widetilde{Y}_2$ are standard generators for the unitary group $\mathcal{C}(\widetilde{H})$ and Y_1, Y_2 are their embeddings into $\mathrm{GL}_{md}(F)$, then $C_{\mathcal{C}(B)}(X) = \langle Y_1, Y_2 \rangle$.

If Q is a quadratic form in even characteristic and $X \in \mathcal{C}(Q)$ is semisimple, then the construction of the generators of $C_{\mathcal{C}(Q)}(X)$ is similar to the sesquilinear case: if $f \in \Phi_1$, then we take standard generators for $\mathcal{C}(Q)$; if $f \in \Phi_2 \cup \Phi_3$, then Proposition 6.1.6 shows that we can repeat the argument of the sesquilinear case by replacing the quadratic form Q with the corresponding bilinear form β_Q.

6.3.1 *Special and Omega Groups*

Let β be a sesquilinear or quadratic form on V, and let $\mathcal{C} = \mathcal{C}(\beta)$ be the isometry group, with corresponding special group $\mathscr{S}$ and Omega group Ω (in the orthogonal case). For a semisimple element X of $\mathscr{S}$ or Ω, we construct a generating set for $C_{\mathcal{C}}(X)$ using the above algorithm. As in the unipotent case discussed in Sect. 4.2, we then use Schreier's algorithm to construct generating sets for $C_{\mathscr{S}}(X)$ and $C_{\Omega}(X)$.

6.4 Constructing a Conjugating Element

Let V be a vector space over $F = \mathbb{F}_{q^u}$, and let $\mathcal{C} = \mathcal{C}(\beta)$ or $\mathcal{C}(Q)$ be the isometry group on V. Given conjugate semisimple elements X and Y of $\mathcal{C}$, we give an algorithm to compute $Z \in \mathcal{C}$ such that $X^Z = Z^{-1}XZ = Y$. It uses Algorithms 1 and 2 from Sect. 6.3. Then we show how to extend the algorithm to address conjugation in $\mathscr{S}$ and Ω.

We first handle the sesquilinear form. Let B be the matrix of a non-degenerate sesquilinear form on V and let $\mathcal{C} = \mathcal{C}(B)$. Let X and Y be conjugate semisimple elements of $\mathcal{C}$. Let $f_1, \dots, f_h \in \Phi$ be the generalized elementary divisors of X and Y, let m_i be their multiplicities and let $d_i = \deg f_i$. The algorithm to compute $Z \in \mathcal{C}$ such that $Z^{-1}XZ = Y$ is the following.

1. Use Algorithm 1 to compute matrices P_X and P_Y in $\mathrm{GL}(V)$ such that $P_X X P_X^{-1} = P_Y Y P_Y^{-1} = J$, where

$$J = \begin{pmatrix} J_1 & & \\ & \ddots & \\ & & J_h \end{pmatrix} \tag{6.4.1}$$

   and J_i is the matrix of the restriction of J to the generalized eigenspace $\ker(f_i(J))$ for every i. Note that J is not necessarily the Jordan form of X: we can choose each J_i completely freely. Hence, J is an isometry for the sesquilinear forms with matrices $B_X = P_X B P_X^*$ and $B_Y = P_Y B P_Y^*$.
2. Compute $W \in C_{\mathrm{GL}(V)}(J)$ such that $W B_Y W^* = B_X$.
3. Let $Z = P_X^{-1} W P_Y$. Now $Z \in \mathcal{C}(B)$ and $Z^{-1}XZ = Y$, so Z is the desired conjugating element.

We now explain Step 2. Algorithm 2 does not return an element of $C_{\mathrm{GL}(V)}(J)$ in general, so more work is needed. By Lemma 6.1.4, the forms B_X and B_Y have block diagonal shape

$$B_X = \begin{pmatrix} B_{X,1} & & \\ & \ddots & \\ & & B_{X,h} \end{pmatrix}, \quad B_Y = \begin{pmatrix} B_{Y,1} & & \\ & \ddots & \\ & & B_{Y,h} \end{pmatrix},$$

and each J_i is an isometry for $B_{X,i}$ and $B_{Y,i}$. So we need to find W_i in the centralizer of J_i such that $W_i B_{Y,i} W_i^* = B_{X,i}$ for every i, and take $W = W_1 \oplus \cdots \oplus W_h$. We distinguish the three cases.

- $f_i \in \Phi_1$. Here J_i is a scalar matrix, so Algorithm 2 returns an element of the centralizer of J_i.
- $f_i \in \Phi_2$, $f_i = g_i g_i{}^*$. If we put

  $$J_i = \begin{pmatrix} \widehat{J_i} & \\ & \widehat{J}_i^* \end{pmatrix},$$

  where $\widehat{J_i}$ is the restriction of J_i to $\ker g_i(J)$, then, by Lemma 6.1.4,

  $$B_{X,i} = \begin{pmatrix} \mathbb{O} & A_{X,i} \\ \varepsilon A_{X,i}^* & \mathbb{O} \end{pmatrix}, \quad B_{Y,i} = \begin{pmatrix} \mathbb{O} & A_{Y,i} \\ \varepsilon A_{Y,i}^* & \mathbb{O} \end{pmatrix},$$

  for some $A_{X,i}$, $A_{Y,i}$ in the centralizer of $\widehat{J_i}$, where $\varepsilon = -1$ if B is alternating, 1 otherwise. We take

  $$W_i = \begin{pmatrix} A_{X,i} A_{Y,i}^{-1} & \mathbb{O} \\ \mathbb{O} & \mathbb{I} \end{pmatrix}.$$

- $f_i \in \Phi_3$. Let $E = F[t]/(f_i)$ in the symplectic and orthogonal cases, $E = \mathbb{F}_q[t]/(f_i \bar{f_i})$ in the unitary case. We follow the argument in the proof of part (iii) of Proposition 6.1.5. Let R be the companion matrix of f_i and let $\varepsilon = -1$ if B is alternating, $\varepsilon = 1$ otherwise. Let J_i be the block diagonal sum of m_i copies of R. Using Algorithm 1, we find T such that $R^* = T^{-1} R^{-1} T$, and, by Claim 1 in the proof of Proposition 6.1.5(iii), we can choose T such that $T = \varepsilon T^*$. Let $\mathcal{T}$ be the block diagonal sum of m_i copies of T. The matrices $H_{X,i} = B_{X,i} \mathcal{T}^{-1}$ and $H_{Y,i} = B_{Y,i} \mathcal{T}^{-1}$ lie in the centralizer of J_i, so they are the embeddings into $\mathrm{GL}_{m_i d_i}(F)$ of matrices $\widetilde{H}_{X,i}$ and $\widetilde{H}_{Y,i}$ in $\mathrm{GL}_{m_i}(E)$. These two matrices are hermitian and, using Algorithm 2, we find $\widetilde{W}_i \in \mathrm{GL}_{m_i}(E)$ such that

  $$\widetilde{H}_{X,i} = \widetilde{W}_i \widetilde{H}_{Y,i} \widetilde{W}_i^\dagger, \tag{6.4.2}$$

where $\widetilde{W}_i \mapsto \widetilde{W}_i^{\dagger}$ is defined as in (6.1.8). By Claim 2 in the proof of Proposition 6.1.5(iii), this implies $B_{X,i} = W_i B_{Y,i} W_i^*$, where W_i is the embedding of $\widetilde{W}_i$ into $\mathrm{GL}_{m_i d_i}(F)$.

This completes the algorithm in the case where $\mathcal{C} = \mathcal{C}(\beta)$.

Now suppose that $\mathcal{C} = \mathcal{C}(Q)$, where Q is a non-degenerate quadratic form. If $f_i \in \Phi_1$, then we apply directly Algorithm 2; if $f_i \in \Phi_2 \cup \Phi_3$, then Proposition 6.1.6 shows that we can replace Q by the corresponding sesquilinear form β_Q and apply the above algorithm.

6.4.1 *Special and Omega Groups*

Let $\mathcal{S}$ and Ω be the special and Omega groups corresponding to the isometry group $\mathcal{C}$. We show how to find conjugating elements in $\mathcal{S}$ and Ω.

In the unitary case, let $X, Y \in \mathcal{S}$ be conjugate in $\mathcal{C}$; by Proposition 6.1.7, they are also conjugate in $\mathcal{S}$. Using the above algorithm we find $Z \in \mathcal{C}$ such that $X^Z = Y$. Let $\lambda = \det(Z)$. As shown in the proof of Proposition 6.1.7(i), we can find an element Z_0 of determinant λ^{-1} in one of the factors of $C_{\mathcal{C}}(X)$. Then ZZ_0 conjugates X to Y and lies in $\mathcal{S}$, as required.

In the orthogonal case, consider first the special group $\mathcal{S}$ with q odd. Let $X, Y \in \mathcal{S}$ be $\mathcal{S}$-conjugate. We compute $Z \in \mathcal{C}$ such that $X^Z = Y$. If $c_S = 2$ (using the notation of Proposition 6.1.7(ii)), then necessarily $Z \in \mathcal{S}$; if $c_S = 1$, then we adjust Z by an element Z_0 of a factor of $C_{\mathcal{C}}(X)$ lying in the appropriate coset of $\mathcal{C} \setminus \mathcal{S}$ such that $ZZ_0 \in \mathcal{S}$.

The case of Ω is handled similarly. Let $X, Y \in \Omega$ be Ω-conjugate. Compute $Z \in \mathcal{S}$ such that $X^Z = Y$. If $c_\Omega = 2$ (using the notation of Proposition 6.1.7(iii)), then necessarily $Z \in \Omega$; if $c_\Omega = 1$, then we adjust Z by an element Z_0 of a factor of $C_{\mathcal{S}}(X)$ lying in the appropriate coset of $\mathcal{S} \setminus \Omega$ such that $ZZ_0 \in \Omega$.

6.5 Some Examples

We close this chapter by listing in Tables 6.1, 6.2, 6.3, and 6.4 respectively the semisimple class representatives and centralizers for the groups $\mathrm{Sp}_6(3)$, $\mathrm{SU}_6(2)$, $\Omega_8^+(2)$ and $\Omega_8^-(2)$. The representatives are identified using the notation of Sect. 6.2. The structures of the centralizers are given by Theorem 6.1.2.

To keep the notation concise, we adopt the following conventions in the tables. In the first column, the superscripts in brackets denote the multiplicities of the elementary divisors; also the elementary divisors in the set Φ_2 are of the form $f = gg^*$, and we give the polynomial g rather than f. In the second column, $f_i \in \Phi_i$ refers to the corresponding polynomial from the first column, and X_{f_i} denotes the representative defined in Sect. 6.2. Again, the superscripts in brackets denote the multiplicities of the diagonal blocks X_{f_i}.

Table 6.1 Semisimple classes for $\mathrm{Sp}_6(3)$

Elementary divisors	Representative	Centralizer
$(t+1)^{[6]}$	$X_{f_1}^{[6]}$	$\mathrm{Sp}_6(3)$
$(t+2)^{[6]}$	$X_{f_1}^{[6]}$	$\mathrm{Sp}_6(3)$
$(t+1)^{[4]}, (t+2)^{[2]}$	$X_{f_1}^{[4]} \oplus X_{f_1}^{[2]}$	$\mathrm{Sp}_4(3) \times \mathrm{Sp}_2(3)$
$(t+2)^{[4]}, (t+1)^{[2]}$	$X_{f_1}^{[4]} \oplus X_{f_1}^{[2]}$	$\mathrm{Sp}_4(3) \times \mathrm{Sp}_2(3)$
$(t+2)^{[4]}, (t^2+1)$	$X_{f_1}^{[4]} \oplus X_{f_3}$	$\mathrm{Sp}_4(3) \times \mathrm{GU}_1(3)$
$(t+1)^{[4]}, (t^2+1)$	$X_{f_1}^{[4]} \oplus X_{f_3}$	$\mathrm{Sp}_4(3) \times \mathrm{GU}_1(3)$
$(t+1)^{[2]}, (t^2+1)^{[2]}$	$X_{f_1}^{[2]} \oplus X_{f_3}^{[2]}$	$\mathrm{Sp}_2(3) \times \mathrm{GU}_2(3)$
$(t+2)^{[2]}, (t^2+1)^{[2]}$	$X_{f_1}^{[2]} \oplus X_{f_3}^{[2]}$	$\mathrm{Sp}_2(3) \times \mathrm{GU}_2(3)$
$(t+1)^{[2]}, (t^2+2t+2)$	$X_{f_1}^{[2]} \oplus X_{f_2}$	$\mathrm{Sp}_2(3) \times \mathrm{GL}_1(3^2)$
$(t+2)^{[2]}, (t^2+2t+2)$	$X_{f_1}^{[2]} \oplus X_{f_2}$	$\mathrm{Sp}_2(3) \times \mathrm{GL}_1(3^2)$
$(t+1)^{[2]}, (t^4+t^3+t^2+t+1)$	$X_{f_1}^{[2]} \oplus X_{f_3}$	$\mathrm{Sp}_2(3) \times \mathrm{GU}_1(3^2)$
$(t+2)^{[2]}, (t^4+t^3+t^2+t+1)$	$X_{f_1}^{[2]} \oplus X_{f_3}$	$\mathrm{Sp}_2(3) \times \mathrm{GU}_1(3^2)$
$(t+1)^{[2]}, (t^4+2t^3+t^2+2t+1)$	$X_{f_1}^{[2]} \oplus X_{f_3}$	$\mathrm{Sp}_2(3) \times \mathrm{GU}_1(3^2)$
$(t+2)^{[2]}, (t^4+2t^3+t^2+2t+1)$	$X_{f_1}^{[2]} \oplus X_{f_3}$	$\mathrm{Sp}_2(3) \times \mathrm{GU}_1(3^2)$
$(t+2)^{[2]}, (t+1)^{[2]}, (t^2+1)$	$X_{f_1}^{[2]} \oplus X_{f_1}^{[2]} \oplus X_{f_3}$	$\mathrm{Sp}_2(3) \times \mathrm{Sp}_2(3) \times \mathrm{GU}_1(3)$
$(t^2+1)^{[3]}$	$X_{f_3}^{[3]}$	$\mathrm{GU}_3(3)$
$(t^2+2t+2), (t^2+1)$	$X_{f_2} \oplus X_{f_3}$	$\mathrm{GL}_1(3^2) \times \mathrm{GU}_1(3)$
$(t^2+1), (t^4+t^3+t^2+t+1)$	$X_{f_3} \oplus X_{f_3}$	$\mathrm{GU}_1(3) \times \mathrm{GU}_1(3^2)$
$(t^2+1), (t^4+2t^3+t^2+2t+1)$	$X_{f_3} \oplus X_{f_3}$	$\mathrm{GU}_1(3) \times \mathrm{GU}_1(3^2)$
t^3+2t+1	X_{f_2}	$\mathrm{GL}_1(3^3)$
t^3+2t^2+2t+2	X_{f_2}	$\mathrm{GL}_1(3^3)$
t^3+2t^2+t+1	X_{f_2}	$\mathrm{GL}_1(3^3)$
t^3+2t+2	X_{f_2}	$\mathrm{GL}_1(3^3)$
$t^6+t^5+t^3+t+1$	X_{f_3}	$\mathrm{GU}_1(3^3)$
$t^6+t^5+t^4+t^3+t^2+t+1$	X_{f_3}	$\mathrm{GU}_1(3^3)$
$t^6+2t^5+2t^3+2t+1$	X_{f_3}	$\mathrm{GU}_1(3^3)$
$t^6+2t^5+t^4+2t^3+t^2+2t+1$	X_{f_3}	$\mathrm{GU}_1(3^3)$

Note that in Table 6.3, there are several classes in $\mathrm{O}_8^+(2)$ that split into two classes in $\Omega_8^+(2)$ (see Proposition 6.1.7(iii)); we have indicated these classes in the first column of the table, and give one representative in the second column. The same applies to Table 6.4.

Finally, in Table 6.2 we denote by ω a primitive element of $\mathbb{F}_4$. In the third column, for convenience, we give the centralizer in $\mathrm{GU}_6(2)$ rather than $\mathrm{SU}_6(2)$; the latter centralizer has index 3 in the given centralizer, by Proposition 6.1.7(i).

Table 6.2 Semisimple classes for $\mathrm{SU}_6(2)$

Elementary divisors	Representative	Centralizer in $\mathrm{GU}_6(2)$
$(t+1)^{[6]}$	$X_{f_1}^{[6]}$	$\mathrm{GU}_6(2)$
$(t+\omega)^{[6]}$	$X_{f_1}^{[6]}$	$\mathrm{GU}_6(2)$
$(t+\omega^2)^{[6]}$	$X_{f_1}^{[6]}$	$\mathrm{GU}_6(2)$
$(t+1)^{[4]}, (t+\omega), (t+\omega^2)$	$X_{f_1}^{[4]} \oplus X_{f_1} \oplus X_{f_1}$	$\mathrm{GU}_4(2) \times \mathrm{GU}_1(2) \times \mathrm{GU}_1(2)$
$(t+1), (t+\omega)^{[4]}, (t+\omega^2)$	$X_{f_1} \oplus X_{f_1}^{[4]} \oplus X_{f_1}$	$\mathrm{GU}_1(2) \times \mathrm{GU}_4(2) \times \mathrm{GU}_1(2)$
$(t+1), (t+\omega), (t+\omega^2)^{[4]}$	$X_{f_1} \oplus X_{f_1} \oplus X_{f_1}^{[4]}$	$\mathrm{GU}_4(2) \times \mathrm{GU}_1(2) \times \mathrm{GU}_1(2)$
$(t+1)^{[3]}, (t+\omega)^{[3]}$	$X_{f_1}^{[3]} \oplus X_{f_1}^{[3]}$	$\mathrm{GU}_3(2) \times \mathrm{GU}_3(2)$
$(t+1)^{[3]}, (t+\omega^2)^{[3]}$	$X_{f_1}^{[3]} \oplus X_{f_1}^{[3]}$	$\mathrm{GU}_3(2) \times \mathrm{GU}_3(2)$
$(t+\omega)^{[3]}, (t+\omega^2)^{[3]}$	$X_{f_1}^{[3]} \oplus X_{f_1}^{[3]}$	$\mathrm{GU}_3(2) \times \mathrm{GU}_3(2)$
$(t+1)^{[2]}, (t+\omega)^{[2]}, (t+\omega^2)^{[2]}$	$X_{f_1}^{[2]} \oplus X_{f_1}^{[2]} \oplus X_{f_1}^{[2]}$	$\mathrm{GU}_2(2) \times \mathrm{GU}_2(2) \times \mathrm{GU}_2(2)$
$(t+1)^{[2]}, (t+\omega), (t^3+\omega^2)$	$X_{f_1}^{[2]} \oplus X_{f_1} \oplus X_{f_3}$	$\mathrm{GU}_2(2) \times \mathrm{GU}_1(2) \times \mathrm{GU}_1(2^3)$
$(t+\omega)^{[2]}, (t+\omega^2), (t^3+\omega^2)$	$X_{f_1}^{[2]} \oplus X_{f_1} \oplus X_{f_3}$	$\mathrm{GU}_2(2) \times \mathrm{GU}_1(2) \times \mathrm{GU}_1(2^3)$
$(t+1), (t+\omega)^{[2]}, (t^3+\omega)$	$X_{f_1} \oplus X_{f_1}^{[2]} \oplus X_{f_3}$	$\mathrm{GU}_1(2) \times \mathrm{GU}_2(2) \times \mathrm{GU}_1(2^3)$
$(t+1)^{[2]}, (t+\omega^2), (t^3+\omega)$	$X_{f_1}^{[2]} \oplus X_{f_1} \oplus X_{f_3}$	$\mathrm{GU}_2(2) \times \mathrm{GU}_1(2) \times \mathrm{GU}_1(2^3)$
$(t+1), (t+\omega^2)^{[2]}, (t^3+\omega^2)$	$X_{f_1} \oplus X_{f_1}^{[2]} \oplus X_{f_3}$	$\mathrm{GU}_1(2) \times \mathrm{GU}_2(2) \times \mathrm{GU}_1(2^3)$
$(t+\omega), (t+\omega^2)^{[2]}, (t^3+\omega)$	$X_{f_1} \oplus X_{f_1}^{[2]} \oplus X_{f_3}$	$\mathrm{GU}_1(2) \times \mathrm{GU}_2(2) \times \mathrm{GU}_1(2^3)$
$(t+1), (t+\omega), (t^2+t+\omega)$	$X_{f_1} \oplus X_{f_1} \oplus X_{f_2}$	$\mathrm{GU}_1(2) \times \mathrm{GU}_1(2) \times \mathrm{GL}_1(2^4)$
$(t+\omega)^{[2]}, (t^2+t+\omega^2)$	$X_{f_1}^{[2]} \oplus X_{f_2}$	$\mathrm{GU}_2(2) \times \mathrm{GL}_1(2^4)$
$(t+\omega), (t+\omega^2), (t^2+\omega t+1)$	$X_{f_1} \oplus X_{f_1} \oplus X_{f_2}$	$\mathrm{GU}_1(2) \times \mathrm{GU}_1(2) \times \mathrm{GL}_1(2^4)$
$(t+1), (t+\omega^2), (t^2+t+\omega^2)$	$X_{f_1} \oplus X_{f_1} \oplus X_{f_2}$	$\mathrm{GU}_1(2) \times \mathrm{GU}_1(2) \times \mathrm{GL}_1(2^4)$
$(t+1)^{[2]}, (t^2+\omega t+1)$	$X_{f_1}^{[2]} \oplus X_{f_2}$	$\mathrm{GU}_2(2) \times \mathrm{GL}_1(2^4)$
$(t+\omega^2)^{[2]}, (t^2+t+\omega)$	$X_{f_1}^{[2]} \oplus X_{f_2}$	$\mathrm{GU}_2(2) \times \mathrm{GL}_1(2^4)$
$(t+\omega), (t^5+\omega^2t^4+\omega^2t^3+t^2+t+\omega^2)$	$X_{f_1} \oplus X_{f_3}$	$\mathrm{GU}_1(2) \times \mathrm{GU}_1(2^5)$
$(t+\omega), (t^5+t^4+\omega^2t^3+t^2+\omega^2t+\omega^2)$	$X_{f_1} \oplus X_{f_3}$	$\mathrm{GU}_1(2) \times \mathrm{GU}_1(2^5)$
$(t+\omega^2), (t^5+\omega t^4+\omega t^3+t^2+t+\omega)$	$X_{f_1} \oplus X_{f_3}$	$\mathrm{GU}_1(2) \times \mathrm{GU}_1(2^5)$
$(t+\omega^2), (t^5+t^4+\omega t^3+t^2+\omega t+\omega)$	$X_{f_1} \oplus X_{f_3}$	$\mathrm{GU}_1(2) \times \mathrm{GU}_1(2^5)$
$(t+1), (t^5+\omega^2t^4+t^3+t^2+\omega t+1)$	$X_{f_1} \oplus X_{f_3}$	$\mathrm{GU}_1(2) \times \mathrm{GU}_1(2^5)$
$(t+1), (t^5+\omega t^4+t^3+t^2+\omega^2t+1)$	$X_{f_1} \oplus X_{f_3}$	$\mathrm{GU}_1(2) \times \mathrm{GU}_1(2^5)$
$t^3+\omega^2t+1$	X_{f_2}	$\mathrm{GL}_1(2^6)$
$t^3+\omega t+1$	X_{f_2}	$\mathrm{GL}_1(2^6)$
t^3+t+1	X_{f_2}	$\mathrm{GL}_1(2^6)$
$(t^3+\omega), (t^3+\omega^2)$	$X_{f_3} \oplus X_{f_3}$	$\mathrm{GU}_1(2^3) \times \mathrm{GU}_1(2^3)$

Table 6.3 Semisimple classes for $\Omega_8^+(2)$

Elementary divisors	Representative	Centralizer
$(t+1)^{[8]}$	$X_{f_1}^{[8]}$	$\Omega_8^+(2)$
$(t+1)^{[6]}$, (t^2+t+1)	$X_{f_1}^{[6]} \oplus X_{f_3}$	$\Omega_6^-(2) \times \mathrm{GU}_1(2)$
$(t+1)^{[4]}$, $(t^2+t+1)^{[2]}$	$X_{f_1}^{[4]} \oplus X_{f_3}^{[2]}$	$\Omega_4^+(2) \times \mathrm{GU}_2(2)$
$(t+1)^{[4]}$, $(t^4+t^3+t^2+t+1)$	$X_{f_1}^{[4]} \oplus X_{f_3}$	$\Omega_4^-(2) \times \mathrm{GU}_1(2^2)$
$(t+1)^{[2]}$, (t^2+t+1), $(t^4+t^3+t^2+t+1)$	$X_{f_1}^{[2]} \oplus X_{f_3} \oplus X_{f_3}$	$\mathrm{GU}_1(2) \times \mathrm{GU}_1(2^2)$
$(t+1)^{[2]}$, $(t^2+t+1)^{[3]}$	$X_{f_1}^{[2]} \oplus X_{f_3}^{[3]}$	$\Omega_2^-(2) \times \mathrm{GU}_3(2)$
$(t+1)^{[2]}$, (t^3+t+1)	$X_{f_1}^{[2]} \oplus X_{f_2}$	$\mathrm{GL}_1(2^3)$
$(t+1)^{[2]}$, (t^6+t^3+1)	$X_{f_1}^{[2]} \oplus X_{f_3}$	$\Omega_2^-(2) \times \mathrm{GU}_1(2^3)$
$(t^2+t+1)^{[4]}$ (two classes)	$X_{f_3}^{[4]}$	$\mathrm{GU}_4(2)$
(t^2+t+1), (t^6+t^3+1) (two classes)	$X_{f_3} \oplus X_{f_3}$	$\mathrm{GU}_1(2) \times \mathrm{GU}_1(2^3)$
$(t^4+t^3+t^2+t+1)^{[2]}$ (two classes)	$X_{f_3}^{[2]}$	$\mathrm{GU}_2(2^2)$
(t^4+t+1) (two classes)	X_{f_2}	$\mathrm{GL}_1(2^4)$

Table 6.4 Semisimple classes for $\Omega_8^-(2)$

Elementary divisors	Representative	Centralizer
$(t+1)^{[8]}$	$X_{f_1}^{[8]}$	$\Omega_8^-(2)$
$(t+1)^{[6]}$, (t^2+t+1)	$X_{f_1}^{[6]} \oplus X_{f_3}$	$\Omega_6^+(2) \times \mathrm{GU}_1(2)$
$(t+1)^{[4]}$, $(t^2+t+1)^{[2]}$	$X_{f_1}^{[4]} \oplus X_{f_3}^{[2]}$	$\Omega_4^-(2) \times \mathrm{GU}_2(2)$
$(t+1)^{[4]}$, $(t^4+t^3+t^2+t+1)$	$X_{f_1}^{[4]} \oplus X_{f_3}$	$\Omega_4^+(2) \times \mathrm{GU}_1(2^2)$
$(t+1)^{[2]}$, $(t^2+t+1)^{[3]}$	$X_{f_1}^{[2]} \oplus X_{f_3}^{[3]}$	$\mathrm{GU}_3(2)$
$(t+1)^{[2]}$, (t^3+t+1)	$X_{f_1}^{[2]} \oplus X_{f_2}$	$\Omega_2^-(2) \times \mathrm{GL}_1(2^3)$
$(t+1)^{[2]}$, (t^6+t^3+1)	$X_{f_1}^{[2]} \oplus X_{f_3}$	$\mathrm{GU}_1(2^3)$
$(t+1)^{[2]}$, (t^2+t+1), $(t^4+t^3+t^2+t+1)$	$X_{f_1}^{[2]} \oplus X_{f_3} \oplus X_{f_3}$	$\Omega_2^-(2) \times \mathrm{GU}_1(2) \times \mathrm{GU}_1(2^2)$
$(t^2+t+1)^{[2]}$, $(t^4+t^3+t^2+t+1)$ (two classes)	$X_{f_3}^{[2]} \oplus X_{f_3}$	$\mathrm{GU}_2(2) \times \mathrm{GU}_1(2^2)$
(t^2+t+1), (t^3+t+1) (two classes)	$X_{f_3} \oplus X_{f_2}$	$\mathrm{GU}_1(2) \times \mathrm{GL}_1(2^3)$
$t^8+t^5+t^4+t^3+1$ (two classes)	X_{f_3}	$\mathrm{GU}_1(2^4)$
$t^8+t^7+t^6+t^4+t^2+t+1$ (two classes)	X_{f_3}	$\mathrm{GU}_1(2^4)$

Chapter 7
General Conjugacy Classes

In this chapter we use the results in the semisimple and unipotent cases to solve the main conjugacy problems (1)–(3) of Sect. 1.1 for classical groups: list representatives for all conjugacy classes; describe the structure of the centralizer of an arbitrary element and provide a generating set; decide conjugacy between arbitrary elements and construct explicit conjugating elements.

7.1 Conjugacy Classes and Centralizers

Let $F = \mathbb{F}_{q^u}$, where $u = 2$ in the unitary case and $u = 1$ otherwise, and let V be an n-dimensional vector space over F. Let $\mathcal{C}$ be the isometry group $\mathcal{C}(\beta)$ or $\mathcal{C}(Q)$, where β is a non-degenerate alternating, symmetric or hermitian form and Q is a non-degenerate quadratic form on V.

Let $x \in \mathcal{C}$. Recall the Jordan decomposition $x = su = us$, with s semisimple and u unipotent. We know that $C_{\mathcal{C}}(x) = C_{\mathcal{C}}(s) \cap C_{\mathcal{C}}(u)$. Moreover, $x_1 = s_1u_1$ and $x_2 = s_2u_2$ are conjugate in $\mathcal{C}$ if and only if s_1 and s_2 are conjugate in $\mathcal{C}$ and $z^{-1}u_1z$ and u_2 are conjugate in $C_{\mathcal{C}}(s_2)$, where $z \in \mathcal{C}$ is such that $z^{-1}s_1z = s_2$. Thus our strategy to list all conjugacy classes of $\mathcal{C}$ is to list all semisimple classes and, for each semisimple representative, list all unipotent classes in its centralizer.

For the semisimple classes of $\mathcal{C}$, we use variants of the representatives produced in Sect. 6.2. Recall the notation of Definition 3.2.4:

$$\Phi_1 = \{f : f \in F[t] \mid f = f^* \text{ monic irreducible, } \deg f = 1\},$$

$$\Phi_2 = \{f : f \in F[t] \mid f = gg^*,\ g \neq g^* \text{ monic irreducible}\},$$

$$\Phi_3 = \{f : f \in F[t] \mid f = f^* \text{ monic irreducible, } \deg f > 1\},$$

G. De Franceschi et al., *Conjugacy in Finite Classical Groups*, Springer Monographs in Mathematics, https://doi.org/10.1007/978-3-031-86461-2_7

and $\Phi = \Phi_1 \cup \Phi_2 \cup \Phi_3$. Each semisimple class of $\mathcal{C}$ can be identified with a pair (S, B), where S is an isometry for the form B. Let $f_1, \ldots, f_h \in \Phi$ be the distinct generalized elementary divisors of S, and let m_i be the multiplicity of f_i for $i = 1, \ldots, h$. Thus, with respect to a suitable basis,

$$S = \begin{pmatrix} S_1 & & \\ & \ddots & \\ & & S_h \end{pmatrix}, \quad B = \begin{pmatrix} B_1 & & \\ & \ddots & \\ & & B_h \end{pmatrix}, \tag{7.1.1}$$

where S_i and B_i are the matrices of the restrictions of S and B to $\ker(f_i(S))$.

We first describe representatives S, then give the corresponding B. For $f_i \in \Phi_1 \cup \Phi_3$, the matrix S_i is a block diagonal sum of m_i companion matrices $C(f_i)$; for $f_i = g_i g_i^* \in \Phi_2$, we take

$$S_i = \begin{pmatrix} Y_i & \\ & Y_i^{*-1} \end{pmatrix},$$

where Y_i is the block diagonal sum of m_i companion matrices $C(g_i)$. For the matrix B, we choose the following form:

- If $f_i \in \Phi_1$, then S_i is a scalar matrix, so B_i can be chosen arbitrarily. Note that $C_{\mathcal{C}(B_i)}(S_i) = \mathcal{C}(B_i)$.
- If $f_i \in \Phi_2$, then choose

$$B_i = \begin{pmatrix} \mathbb{O} & \mathbb{I} \\ \varepsilon\mathbb{I} & \mathbb{O} \end{pmatrix}, \quad \varepsilon = \begin{cases} 1 & \text{if } B \text{ is hermitian or symmetric;} \\ -1 & \text{if } B \text{ is alternating;} \\ 0 & \text{if } B \text{ is quadratic.} \end{cases}$$

  Note that $C_{\mathcal{C}(B_i)}(S_i) \cong \mathrm{GL}_{m_i}(q^{d_i})$, where $d_i = u \deg(f_i)/2$ (see Theorem 6.1.2).
- If $f_i \in \Phi_3$, then choose

$$B_i = \begin{pmatrix} & & B_{f_i} \\ & \cdot^{\cdot^{\cdot}} & \\ B_{f_i} & & \end{pmatrix}, \tag{7.1.2}$$

  where B_{f_i} is the matrix of a form preserved by the companion matrix $C(f_i)$, computed as discussed at the end of Sect. 6.2, and it appears m_i times. Note that $C_{\mathcal{C}(B_i)}(S_i) \cong \mathrm{GU}_{m_i}(q^{d_i})$ where $d_i = u \deg(f_i)/2$ (see Theorem 6.1.2).

Let $x = su = us \in \mathcal{C}$, with s semisimple and u unipotent. Choose a basis such that s and the form (β or Q) have matrices S and B respectively as in (7.1.1). Let U be the matrix of u with respect to this basis. We know that U belongs to

$C_{\mathcal{C}}(S) = \prod_{i=1}^{h} C_{\mathcal{C}(B_i)}(S_i)$. Thus U is a block diagonal matrix

$$\begin{pmatrix} U_1 & & \\ & \ddots & \\ & & U_h \end{pmatrix}$$

where U_i is the matrix of the restriction of u to $\ker(f_i(S))$.

With the above notation, we can now describe the conjugacy classes and centralizers in $\mathcal{C}$. The result follows from the above discussion.

Theorem 7.1.1 *A complete set of representatives for the conjugacy classes of $\mathcal{C}$ is given by all pairs of matrices (SU, B), defined by*

$$S = \begin{pmatrix} S_1 & & \\ & \ddots & \\ & & S_h \end{pmatrix}, \quad U = \begin{pmatrix} U_{1j} & & \\ & \ddots & \\ & & U_{hj} \end{pmatrix}, \quad B = \begin{pmatrix} B_1 & & \\ & \ddots & \\ & & B_h \end{pmatrix}.$$

where (S, B) runs over all representatives of semisimple conjugacy classes of $\mathcal{C}$ and, for each i, the matrices U_{ij} run over all representatives of unipotent conjugacy classes of $\mathcal{C}_i := C_{\mathcal{C}(B_i)}(S_i)$. For a given representative SU as above,

$$C_{\mathcal{C}}(SU) = \prod_{i=1}^{h} C_{\mathcal{C}_i}(U_{ij}).$$

The factors $C_{\mathcal{C}_i}(U_{ij})$ are described in Sect. 2.2.2 when $\mathcal{C}_i = \mathrm{GL}_{m_i}(q^{d_i})$; Sects. 4.2.1 and 5.2 when $\mathcal{C}_i$ is symplectic; Sects. 4.2.2 and 5.2 when $\mathcal{C}_i$ is orthogonal; and Sect. 4.2.3 when $\mathcal{C}_i$ is unitary.

7.1.1 Listing Class Representatives in the Isometry Group

In Sect. 6.2 we showed how to list the semisimple class representatives in $\mathcal{C}$. It remains to list representatives for all conjugacy classes of $\mathcal{C}$ with fixed semisimple part S. With notation as above, this amounts to listing representatives for all unipotent classes of $C_{\mathcal{C}(B_i)}(S_i)$. We distinguish the three cases.

- $f_i \in \Phi_1$. Here S_i is a scalar matrix and $C_{\mathcal{C}(B_i)}(S_i)$ coincides with $\mathcal{C}(B_i)$. The representatives for unipotent classes of $\mathcal{C}(B_i)$ are given in Sects. 4.1 and 5.1, and we are free to choose the form B_i.

- $f_i \in \Phi_2$, $f_i = g_i g_i{}^*$. Now $C_{\mathcal{C}(B_i)}(S_i)$ is isomorphic to $\mathrm{GL}_{m_i}(E)$, with $E = F[t]/(g_i)$, via the isomorphism

$$y \mapsto \begin{pmatrix} Y & \\ & Y^{*-1} \end{pmatrix}, \ \forall y \in \mathrm{GL}_{m_i}(E),$$

where Y is the embedding of y into $\mathrm{GL}_{m_i d_i}(F)$. Two elements of $\mathrm{GL}_{m_i}(E)$ are conjugate if and only if they have the same generalized elementary divisors, so the list of representatives of unipotent classes of $C_{\mathcal{C}(B_i)}(S_i)$ is just the list of the images in $C_{\mathcal{C}(B_i)}(S_i)$ of representatives of the unipotent classes of $\mathrm{GL}_{m_i}(E)$. We choose the block diagonal sum of unipotent Jordan blocks as our preferred form.
- $f_i \in \Phi_3$. Let $E = F[t]/(f_i)$ in the symplectic and orthogonal cases, and $E = \mathbb{F}_q[t]/(f_i \bar{f_i})$ in the unitary case. By Proposition 6.1.5 and its proof, $C_{\mathcal{C}(B_i)}(S_i)$ is the embedding in $\mathrm{GL}_{m_i d_i}(F)$ of the group $\mathrm{GU}_{m_i}(E)$ preserving the hermitian form with matrix

$$\begin{pmatrix} & & 1 \\ & \cdot^{\cdot^{\cdot}} & \\ 1 & & \end{pmatrix}.$$

This follows from our choice of B_i in (7.1.2). Representatives for all unipotent classes of $\mathrm{GU}_{m_i}(E)$ are the Jordan forms $\bigoplus_{i=1}^{k} V(m_i)^{[r_i]}$ given in Sect. 4.1.3.

7.1.2 *Listing Class Representatives in Special and Omega Groups*

We now determine the splitting of those classes in the isometry groups $\mathcal{C} = \mathrm{GU}_n(q)$ or $\mathrm{O}_n^{\epsilon}(q)$ that lie in the special groups $\mathscr{S} = \mathrm{SU}_n(q)$ or $\mathrm{SO}_n^{\epsilon}(q)$, or in the Omega group $\Omega = \Omega_n^{\epsilon}(q)$. This is a matter of computing the indices $|C_{\mathcal{C}}(X) : C_{\mathscr{S}}(X)|$ and $|C_{\mathcal{C}}(X) : C_{\Omega}(X)|$ for elements X of $\mathscr{S}$ and Ω.

Proposition 7.1.2

(i) *Let $\mathcal{C} = \mathrm{GU}_n(q)$ and $\mathscr{S} = \mathrm{SU}_n(q)$. Let $X \in \mathscr{S}$ have generalized elementary divisors $f_1^{m_1}, \ldots, f_k^{m_k}$, where the $f_i \in \Phi$ are not necessarily distinct. Then $|C_{\mathcal{C}}(X) : C_{\mathscr{S}}(X)| = (q+1)/r$, where $r = \gcd(m_1, \ldots, m_k, q+1)$. The class $X^{\mathcal{C}}$ splits into r classes in $\mathscr{S}$.*

(ii) *Let $\mathcal{C} = \mathrm{O}_n^{\epsilon}(q)$ and $\mathscr{S} = \mathrm{SO}_n^{\epsilon}(q)$ with q odd, and let $X \in \mathscr{S}$. Then $|C_{\mathcal{C}}(X) : C_{\mathscr{S}}(X)| = c_S$, where*

$$c_S = \begin{cases} 2, & \text{if } X \text{ has an elementary divisor } (t \pm 1)^m \text{ with } m \text{ odd}, \\ 1, & \text{otherwise.} \end{cases}$$

The class $X^{\mathcal{C}}$ splits into $2/c_S$ classes in $\mathscr{S}$.

Proof

(i) This is similar to the proof of Theorem 2.3.1. As in that proof, we can reduce to the case where the generalized elementary divisors are $f^{m_1}, \dots, f^{m_k}$, all powers of a given polynomial $f \in \Phi$. Let $X = SU = US$ be the Jordan decomposition of X, and let $D := \{\lambda \in \mathbb{F}_{q^2} : \lambda^{q+1} = 1\} = \langle \omega \rangle$.

Suppose first that $f \in \Phi_1$, so that $f(t) = t - \lambda$ with $\lambda \in D$. Then $C_{\mathcal{C}}(S) = \mathcal{C}$ and $U \in \mathcal{C}$ is unipotent with Jordan form $\bigoplus_{i=1}^{l} J_{m_i}$. Relabel the m_i so that this Jordan form is $\bigoplus_{i=1}^{k} J_{\lambda_i}^{[l_i]}$, where $\lambda_1, \dots, \lambda_k$ are distinct. By Theorem 4.1.4, $C_{\mathcal{C}}(U) = Q_0 R$, where Q_0 is a normal subgroup of q-power order and $R \cong \prod_{i=1}^{k} \mathrm{GU}_{l_i}(q)$. The action of R on $V = V_n(q^2)$ is as $\prod_{i=1}^{k}(\mathrm{GU}_{l_i}(q) \otimes I_{\lambda_i})$ (since there is such a subgroup in $C_{\mathcal{C}}(U)$). If $A \in \mathrm{GU}_{l_i}(q)$ has determinant ω, then the determinant of $A \otimes I_{\lambda_i}$ is ω^{λ_i}. Hence the subgroup of $\mathbb{F}_{q^2}^*$ consisting of all determinants of elements of R is generated by ω^{λ_i} for $i = 1, \dots, k$. This equals $\langle \omega^r \rangle$ where $r = \gcd(\lambda_1, \dots, \lambda_k, q+1)$. Hence the image of the determinant map $C_{\mathrm{GU}_n(q)}(X) \to D$ has order $(q+1)/r$, completing the proof for the case $f \in \Phi_1$.

Next suppose $f \in \Phi_2$, so that $f = gg^*$ with $g \neq g^*$ irreducible of degree d. By Theorem 6.1.2, $C_{\mathcal{C}}(S) \cong \mathrm{GL}_m(q^{2d})$, where $m = \sum m_i$. Relabel the m_i so that the Jordan form of the unipotent element U of $\mathrm{GL}_m(q^{2d})$ is $\bigoplus_{i=1}^{k} J_{\lambda_i}^{[l_i]}$, where $\lambda_1, \dots, \lambda_k$ are distinct. By Theorem 2.2.3, $C_{\mathcal{C}}(X) = C_{\mathrm{GL}_m(q^{2d})}(U) = Q_0 R$, where $R \cong \prod_{i=1}^{k} \mathrm{GL}_{l_i}(q^{2d})$ acts on $V = V_m(q^{2d})$ as $\prod_{i=1}^{k}(\mathrm{GL}_{l_i}(q^{2d}) \otimes I_{\lambda_i})$. There exists $A \in \mathrm{GL}_{l_i}(q^{2d})$ whose determinant as an element of $\mathrm{GU}_{dl_i}(q)$ is ω, and so the determinant of $A \otimes I_{\lambda_i}$ is ω^{λ_i}. Now we complete the argument as before.

The final case, where $f \in \Phi_3$, is similar. Here f is irreducible of odd degree $d > 1$, and $C_{\mathcal{C}}(S) \cong \mathrm{GU}_m(q^d)$ by Theorem 6.1.2. Again relabel the m_i so that the Jordan form of $U \in \mathrm{GU}_m(q^d)$ is $\bigoplus_{i=1}^{k} J_{\lambda_i}^{[l_i]}$, where $\lambda_1, \dots, \lambda_k$ are distinct. Then, by Theorem 4.1.4, $C_{\mathcal{C}}(X) = C_{\mathrm{GU}_m(q^d)}(U) = Q_0 R$, where $R \cong \prod_{i=1}^{k} \mathrm{GU}_{l_i}(q^d)$ acts on $V = V_m(q^{2d})$ as $\prod_{i=1}^{k}(\mathrm{GU}_{l_i}(q^d) \otimes I_{\lambda_i})$. Now we complete the proof as before.

(ii) Let $X = SU$ be the Jordan decomposition of X. For every generalized elementary divisor $f_i \in \Phi_2 \cup \Phi_3$, the corresponding factor of $C_{\mathcal{C}}(S)$ lies in $\mathscr{S}$ by Proposition 6.1.7(ii). So we may suppose that each generalized elementary divisor $f_i^{m_i}$ is $(t-1)^{m_i}$ or $(t+1)^{m_i}$. If all the m_i are even, then

$$C_{\mathcal{C}}(X) \leq C_{\mathrm{GL}(V)}(X) \leq \mathrm{SL}(V)$$

(see Sect. 2.3). On the other hand, if there is a generalized elementary divisor $(t \pm 1)^m$ with m odd, then, with respect to a suitable basis, $C_{\mathcal{C}}(X)$ contains a block diagonal matrix $-I_m \oplus I_{n-m}$, which has determinant -1. □

The splitting of classes in $\Omega = \Omega_n^\epsilon(q)$ is more complicated, and is discussed in the next two propositions. Let $X \in \Omega$ have Jordan decomposition SU with S semisimple and U unipotent. Let $f_1, \ldots, f_h$ be the generalized elementary divisors of S, with multiplicities $m_1, \ldots, m_h$. Change the basis so that X and the form B are

$$X = \begin{pmatrix} X_1 & & \\ & \ddots & \\ & & X_h \end{pmatrix}, \quad B = \begin{pmatrix} B_1 & & \\ & \ddots & \\ & & B_h \end{pmatrix},$$

where X_i and B_i are the matrices of the restrictions of X and B respectively to $\ker(f_i(S))$. Write $X_i = S_i U_i$, where S_i and U_i are the semisimple and unipotent parts respectively. Abbreviate $\mathcal{C}(B)$, $\mathcal{S}(B)$ and $\Omega(B)$ by $\mathcal{C}$, $\mathcal{S}$ and Ω respectively.

Proposition 7.1.3 *Let $X \in \Omega$ be as above, and assume q is odd. Then $|C_{\mathcal{S}}(X) : C_\Omega(X)| = c_\Omega$, where $c_\Omega \in \{1, 2\}$, and $c_\Omega = 1$ if and only if the following conditions hold:*

(i) *for a generalized elementary divisor $t \pm 1$, the corresponding unipotent element U_i takes the following shape (following the notation of (4.1.5)):*

$$U_i = \bigoplus_{j=1}^{r} V_{\beta_j}(2k_j + 1) \oplus \bigoplus_{j=1}^{s} W(2l_j)^{[b_j]},$$

where either $r = 0$, or $r \geq 1$, the integers $k_1, \ldots, k_r$ are distinct, and the $\beta_j(-1)^{k_j}$ are mutually congruent modulo $(\mathbb{F}_q^*)^2$;

(ii) *if both $t - 1$ and $t + 1$ are generalized elementary divisors of X, then the corresponding unipotent elements U_i take the following shape:*

$$U_1 = \bigoplus_{j=1}^{r} V_{\beta_j}(2k_j + 1) \oplus \sum_{j=1}^{s} W(2l_j)^{[b_j]},$$

$$U_2 = \bigoplus_{j=1}^{r'} V_{\beta'_j}(2k'_j + 1) \oplus \bigoplus_{j=1}^{s'} W(2l'_j)^{[b'_j]} \tag{7.1.3}$$

(as in (i)), and all of the quantities $\beta_j(-1)^{k_j}$ and $\beta'_j(-1)^{k'_j}$ are mutually congruent modulo $(\mathbb{F}_q^*)^2$;

(iii) *X has no generalized elementary divisors f^k with $f \in \Phi_2 \cup \Phi_3$ and k odd.*

The class $X^{\mathcal{S}}$ splits into $2/c_\Omega$ conjugacy classes in Ω.

Proof Suppose $c_\Omega = 1$. We show that (i)–(iii) hold.

Consider a generalized elementary divisor $f_i \in \Phi_2$ of S. The centralizer $C_{\mathcal{C}(B_i)}(S_i)$ is isomorphic to $\mathrm{GL}_{m_i}(q^{d_i})$ (where $d_i = \deg(f_i)/2$). As shown in the proof of Proposition 6.1.7, it intersects $\Omega(B_i)$ in the subgroup of index 2 consisting of elements whose determinant is a square in $\mathbb{F}^*_{q^{d_i}}$. The centralizer of the unipotent element U_i in $\mathrm{GL}_{m_i}(q^{d_i})$ contains an element of non-square determinant if and only if it has a block of odd size k. As $c_\Omega = 1$, this cannot be the case, so (iii) holds. (This also shows that if (iii) fails, then $c_\Omega = 2$.) A similar argument applies for a generalized elementary divisor $f_i \in \Phi_3$.

Now consider a generalized elementary divisor $f_i = t \pm 1$. Here $S_i = \pm I$ and U_i, the corresponding unipotent element, is

$$\bigoplus_{j=1}^{r}(V_{\beta_j}(2k_j+1) \oplus V_1(2k_j+1)^{[a_j-1]}) \oplus \bigoplus_{j=1}^{s} W(2l_j)^{[b_j]},$$

in the notation of (4.1.5). As shown in the proof of [14, Prop. 2.4(iii)], $C_{C(B_i)}(U_i)$ is contained in $\Omega(B_i)$ if and only if condition (i) holds for U_i. In particular, if $c_\Omega = 1$, then (i) holds.

Suppose that both $t-1$ and $t+1$ are generalized elementary divisors of X. Since $c_\Omega = 1$, it follows that U_1 and U_2 satisfy condition (i). Suppose U_1 and U_2 are as in (7.1.3). If for some a, b the quantities $\beta_a(-1)^{k_a}$ and $\beta'_b(-1)^{k'_b}$ are not congruent modulo $(\mathbb{F}_q^*)^2$, then $-I^T$, where $T = V_{2k_a+1} \oplus V_{2k'_b+1}$, centralizes X and lies in $C \setminus \Omega$ (see the proof of [14, Prop. 2.4(iii)]). Hence, if $c_\Omega = 1$, then condition (ii) must hold.

We have shown that if $c_\Omega = 1$, then conditions (i)–(iii) hold. The converse follows from various observations made in the proof. □

Proposition 7.1.4 *Let $X \in \Omega$ be as above, and assume q is even. Then $|C_C(X) : C_\Omega(X)| = c_\Omega$, where $c_\Omega \in \{1, 2\}$, and $c_\Omega = 1$ if and only if one of the following conditions holds:*

(i) *there is no generalized elementary divisor $f_i = t + 1$;*
(ii) *there is a generalized elementary divisor $f_i = t + 1$, and the corresponding unipotent element U_i is in the class $\bigoplus_{j=1}^{s} W(2l_j)^{[b_j]}$ (following the notation of (5.1.1)).*

Proof By the proof of Proposition 6.1.7(iii), if X has a generalized elementary divisor $f_i \in \Phi_2 \cup \Phi_3$, then $C_{C(B_i)}(X_i) \leq C_{C(B_i)}(S_i) \leq \Omega(B_i)$. So we only need to consider the case where X has a generalized elementary divisor $f_i = t + 1$; now $X_i = U_i$ is a unipotent element as in (5.1.1). Moreover, as observed after (5.1.1), $C_{C(B_i)}(X_i) \leq \Omega(B_i)$ if and only $U_i = \bigoplus_{j=1}^{s} W(2l_j)^{[b_j]}$, as in part (ii). □

7.2 Generators for the Centralizer of a General Element

Having described the centralizer of an arbitrary element of a classical group in Sect. 7.1, we now show how to construct a generating set for it. We use Algorithms 1 and 2 described in Sect. 6.3, and the algorithms of Sects. 4.2 and 5.2 which solve the following problem:

- (*Algorithm* 3) Let $\mathcal{C} = \mathcal{C}(\beta)$ or $\mathcal{C}(Q)$ be a classical isometry group. Given unipotent $X \in \mathcal{C}$, return a generating set for $C_{\mathcal{C}}(X)$.

Let B be the matrix of a non-degenerate sesquilinear form on V. Let $X \in \mathcal{C}(B)$, and let the generalized elementary divisors of X be

$$f_i^{m_{ij}} \quad (i = 1, \ldots, h,\ 1 \le j \le k_i),$$

where $f_1, \ldots, f_h \in \Phi$ are distinct, and $m_{i1} \ge m_{i2} \ge \cdots \ge m_{ik_i}$ for all i. We change basis so that

$$X = \begin{pmatrix} X_1 & & \\ & \ddots & \\ & & X_h \end{pmatrix}, \quad B = \begin{pmatrix} B_1 & & \\ & \ddots & \\ & & B_h \end{pmatrix},$$

where X_i and B_i are the matrices of the restrictions of X and B to $\ker(f_i(X)^{m_{i1}})$. Let $m_i = \sum_{j=1}^{k_i} m_{ij}$ for each i. We can suppose that X_i is in Jordan form if $f_i \in \Phi_1 \cup \Phi_3$, or

$$X_i = \begin{pmatrix} \widehat{X}_i & \\ & \widehat{X}_i^{*-1} \end{pmatrix} \tag{7.2.1}$$

if $f_i = g_i g_i{}^* \in \Phi_2$, where $\widehat{X}_i$ is the Jordan form of the matrix of the restriction of X to $\ker(g_i(X)^{m_{i1}})$. Let

$$E_i = \begin{cases} F & \text{if } f_i \in \Phi_1 \\ F[t]/(g_i) & \text{if } f_i \in \Phi_2,\ f_i = g_i g_i{}^* \\ F[t]/(f_i) & \text{if } f_i \in \Phi_3, \text{ symplectic and orthogonal cases} \\ \mathbb{F}_q[t]/(f_i \bar{f}_i) & \text{if } f_i \in \Phi_3, \text{ unitary case.} \end{cases}$$

Finally, let

$$\begin{pmatrix} X_1 & & \\ & \ddots & \\ & & X_h \end{pmatrix} = \begin{pmatrix} S_1 & & \\ & \ddots & \\ & & S_h \end{pmatrix} \begin{pmatrix} U_1 & & \\ & \ddots & \\ & & U_h \end{pmatrix}$$

be the Jordan decomposition of X, and

$$\begin{pmatrix} \widehat{X}_i & \\ & \widehat{X}_i^{*-1} \end{pmatrix} = \begin{pmatrix} \widehat{S}_i & \\ & \widehat{S}_i^{*-1} \end{pmatrix} \begin{pmatrix} \widehat{U}_i & \\ & \widehat{U}_i^{*-1} \end{pmatrix}$$

when $f_i \in \Phi_2$. A generating set for $C_{\mathcal{C}}(X)$ consists of the matrices

$$y_{ij} = \begin{pmatrix} \mathbb{I} & & \\ & Y_{ij} & \\ & & \mathbb{I} \end{pmatrix}, \tag{7.2.2}$$

where the $\mathbb{I}$'s are identity matrices of the appropriate sizes, and Y_{ij} runs over a generating set for $C_{\mathcal{C}(B_i)}(X_i)$. These are obtained as follows.

- $f_i \in \Phi_1$. The Y_{ij} are the generators for $C_{\mathcal{C}(B_i)}(U_i)$ returned by Algorithm 3.
- $f_i \in \Phi_2$. Let $d_i' = \deg(f_i)/2$. By Lemma 6.1.4, the form preserved by X_i is

  $$B_i = \begin{pmatrix} \mathbb{O} & A_i \\ \varepsilon A_i^* & \mathbb{O} \end{pmatrix},$$

  where $\varepsilon = -1$ in the symplectic case and 1 otherwise. Now $\widehat{U}_i$ is the embedding of a unipotent $\widetilde{U}_i \in \mathrm{GL}_{m_i}(E_i)$. We take

  $$Y_{ij} = \begin{pmatrix} Z_{ij} & \\ & A_i^* Z_{ij}^{*-1} A_i^{*-1} \end{pmatrix}, \tag{7.2.3}$$

  where the Z_{ij} are the embeddings into $\mathrm{GL}_{m_i d_i'}(F)$ of the generators of $C_{\mathrm{GL}_{m_i}(E_i)}(\widetilde{U}_i)$, as described in Sect. 2.2.2.
- $f_i \in \Phi_3$. Define E_i as above. Let R be the companion matrix $C(f_i)$ and let $\varepsilon = -1$ if B is alternating, $\varepsilon = 1$ otherwise. We can suppose that S_i is the block diagonal sum of m_i copies of R. Using Algorithm 1 we find T such that $R^* = T^{-1}R^{-1}T$. By Claim 1 in the proof of Proposition 6.1.5(iii), we can choose T such that $T = \varepsilon T^*$. Let $\mathcal{T}$ be the block diagonal sum of m_i copies of T. The matrix $H_i = B_i\mathcal{T}^{-1}$ lies in the centralizer of S_i, so it is the embedding into $\mathrm{GL}_{m_i d_i}(F)$ of $\widetilde{H}_i \in \mathrm{GL}_{m_i}(E_i)$. Claim 2 in the proof of Proposition 6.1.5(iii) shows that $\widetilde{H}_i$ is hermitian and U_i is the embedding into $\mathrm{GL}_{m_i d_i}(F)$ of a unipotent $\widetilde{U}_i \in \mathcal{C}(\widetilde{H}_i) \cong \mathrm{GU}_{m_i}(E_i)$. So $C_{\mathcal{C}(B_i)}(X_i)$ is generated by the embeddings into $\mathrm{GL}_{m_i d_i}(F)$ of the generators of $C_{\mathcal{C}(\widetilde{H}_i)}(\widetilde{U}_i)$ returned by Algorithm 3.

This addresses the case where the classical group $\mathcal{C} = \mathcal{C}(B)$. For the case $\mathcal{C} = \mathcal{C}(Q)$, an orthogonal group in characteristic 2, note that in the analysis of the cases $f_i \in \Phi_2 \cup \Phi_3$, if Q_i is a quadratic form, then it can be replaced by the associated bilinear form β_{Q_i}, so we can proceed as above.

7.2.1 Special and Omega Groups

To obtain generators for the centralizer in $\mathscr{S}$ of $X \in \mathscr{S}$, we apply Schreier's algorithm to $C_C(X)$. If $X \in \Omega$, then we apply the algorithm to $C_{\mathscr{S}}(X)$.

7.3 The Conjugacy Problem

Having solved the conjugacy problem for unipotent elements in Sects. 4.3 and 5.3, and for semisimple elements in Sect. 6.1, we now solve the problem in general.

Theorem 7.3.1 *Let C be a classical group on $V = F^n$, where $F = \mathbb{F}_{q^u}$ (with $u = 2$ in the unitary case and $u = 1$ otherwise). Let $X, Y \in C$, let $\prod_{i=1}^{h} f_i^{m_i}$ be the minimal polynomial of X, with $f_1, \ldots, f_h \in \Phi$ distinct, and let X_i be the restriction of X to $\ker(f_i(X)^{m_i})$, with similar notation for Y, Y_i.*

(i) *If $C = \mathrm{GU}_n(q)$, then X and Y are conjugate in C if and only if $X \sim Y$.*
(ii) *If $C = \mathrm{Sp}_n(q)$ or $\mathrm{O}_n^{\epsilon}(q)$, then X and Y are conjugate in C if and only if $X \sim Y$ and, for every i such that $f_i(t) = t \pm 1$, the unipotent parts of X_i and Y_i are conjugate in the corresponding symplectic or orthogonal group (see Sects. 4.3 and 5.3).*
(iii) *If $\mathscr{S} = \mathrm{SU}_n(q)$ and $X, Y \in \mathscr{S}$, then X and Y are conjugate in $\mathscr{S}$ if and only if X and Y are conjugate in $\mathrm{SL}(V)$: namely, X and Y are conjugate in C and every conjugating element in C has determinant a power of ω^r, where $\omega \in F$ has order $q + 1$ and r is the greatest common divisor of $q + 1$ and the dimensions of the Jordan blocks of X.*
(iv) *If $\mathscr{S} = \mathrm{SO}_n^{\epsilon}(q)$ with q odd, and $X, Y \in \mathscr{S}$, then X and Y are conjugate in $\mathscr{S}$ if and only if X and Y are conjugate in C, and either each has an elementary divisor $(t \pm 1)^e$ with e odd, or every conjugating element in C has determinant 1.*
(v) *If $\Omega = \Omega_n^{\epsilon}(q)$, and $X, Y \in \Omega$, then X and Y are conjugate in Ω if and only if they are conjugate in $\mathscr{S}$, and either their class in $\mathscr{S}$ does not split into two distinct classes in Ω (see Propositions 7.1.3 and 7.1.4), or every conjugating element in $\mathscr{S}$ lies in Ω.*

Note that for the criteria in parts (iii)–(v), conjugating elements in the relevant groups are constructed in the next section.

Proof

(i) The left-to-right implication is obvious, so consider the converse. Let $X, Y \in C = \mathrm{GU}(V)$ and suppose $X \sim Y$. Write $X = SU$, $Y = S'U'$ for the Jordan decompositions. Then $S \sim S'$, so S is conjugate to S' in C by Theorem 6.1.1. Hence, replacing Y by a conjugate, we can take $S = S'$, so U and U' are similar unipotent elements in $C_C(S)$. By Theorem 6.1.2, $C_C(S)$ is a direct product of general linear and general unitary groups, and so U and U' are conjugate in $C_{\mathcal{C}}(S)$. Thus X and Y are conjugate in $\mathcal{C}$, as required.

(ii) Again, we only need to prove the right-to-left implication. Assume that $X \sim Y$ and that for $f_i = t \pm 1$, the unipotent parts of X_i and Y_i are conjugate in the corresponding symplectic or orthogonal group. Write $X = SU$, $Y = S'U'$ for the Jordan decompositions. Then S is conjugate to S' in $\mathcal{C}$ by Theorem 6.1.1, so we can take $S = S'$. For $f_i = t \pm 1$ write $V_i = \ker(f_i(S))$, so by Theorem 6.1.2

$$C_{\mathcal{C}}(S) = \prod_{f_i = t\pm 1} \mathcal{C}(V_i) \times D,$$

where D is a direct product of general linear and general unitary groups. Then U and U' are unipotent elements of this centralizer, and by assumption their projections to the first factor are conjugate in that factor. It follows that U and U' are conjugate in $C_{\mathcal{C}}(S)$, giving the conclusion.

(iii) Suppose $X, Y \in \mathcal{S}$ are conjugate in $\mathcal{S}$, say $Y = X^Z$ with $Z \in \mathcal{S}$. If $Y = X^{Z'}$ with $Z' \in \mathcal{C}$, then $Z'Z^{-1} \in C_{\mathcal{C}}(X)$, and so $\det(Z')$ is a power of ω^r, as shown in the proof of Proposition 7.1.2(i).

For the converse, suppose that $X, Y \in \mathcal{S}$ and $Y = X^Z$, where $Z \in \mathcal{C} = \mathrm{GU}(V)$ has determinant a power of ω^r. As before, write $X = SU$ and $Y = S'U'$ for the Jordan decompositions. Then S is conjugate to S' in $\mathcal{S}$ by Proposition 6.1.7, so we can take $S = S'$. Now the proof of Proposition 7.1.2(i) shows that U and U' are conjugate in $C_{\mathcal{S}}(S)$, as required.

(iv) Let $X, Y \in \mathcal{S}$ be conjugate in $\mathcal{C}$. In the notation of Proposition 7.1.2(ii), if $c_S = 2$, then $X^{\mathcal{C}} = X^{\mathcal{S}}$ and so X and Y are conjugate in $\mathcal{S}$; and $c_S = 1$ if and only if X has an elementary divisor $(t \pm 1)^e$ with e odd. Part (iv) follows.

(v) Let $X, Y \in \Omega$ be conjugate in $\mathcal{S}$. We use the notation of Propositions 7.1.3 and 7.1.4. If $c_\Omega = 2$, then $X^\Omega = X^{\mathcal{S}}$ and so X and Y are conjugate in Ω; if $c_\Omega = 1$, then $C_\Omega(X) = C_{\mathcal{S}}(X)$, and so every conjugating element in $\mathcal{S}$ lies in Ω. This completes the proof. □

7.4 Constructing a Conjugating Element

Let $\mathcal{C} = \mathcal{C}(B)$ or $\mathcal{C}(Q)$ be a classical isometry group. Given conjugate $X, Y \in \mathcal{C}$, we describe an algorithm to compute $Z \in \mathcal{C}$ such that $Y = X^Z$. We use Algorithms 1 and 2 described in Sect. 6.3, together with the algorithms of Sects. 4.4 and 5.4 which solve the following problem.

- (*Algorithm* 4) Given unipotent $X, Y \in \mathcal{C}$ that are conjugate in $\mathcal{C}$, return $Z \in \mathcal{C}$ such that $Y = X^Z$.

The algorithm is similar to the semisimple case, described in Sect. 6.4. Consider first $\mathcal{C} = \mathcal{C}(B)$. The procedure begins as in Step 1 of Sect. 6.4, by computing matrices P_X and P_Y such that $P_X X P_X^{-1} = P_Y Y P_Y^{-1} = J$, where J is as in (6.4.1) and the J_i are the restrictions of J to the generalized eigenspaces. Then J preserves the forms $B_X = P_X B P_X^*$ and $B_Y = P_Y B P_Y^*$, where

$$J = \begin{pmatrix} J_1 & & \\ & \ddots & \\ & & J_h \end{pmatrix}, \quad B_X = \begin{pmatrix} B_{X,1} & & \\ & \ddots & \\ & & B_{X,h} \end{pmatrix}, \quad B_Y = \begin{pmatrix} B_{Y,1} & & \\ & \ddots & \\ & & B_{Y,h} \end{pmatrix}$$

and each J_i is an isometry for $B_{X,i}$ and $B_{Y,i}$. As in Sect. 6.4, Step 2 of the algorithm is to compute W_i in the centralizer of J_i such that $W_i B_{Y,i} W_i^* = B_{X,i}$ for each i. Taking $W = \bigoplus_{i=1}^h W_i$ and $Z = P_X^{-1} W P_Y$, we deduce that $Z \in \mathcal{C}(B)$ and $X^Z = Y$, as required.

To compute the W_i, we distinguish the three cases.

- $f_i \in \Phi_1$. Now J_i is a product of a scalar and a unipotent element, and we can suppose that the scalar is the identity (so J_i is unipotent) because it does not affect the computation. Using Algorithm 2, we compute $W_{B,i}$ such that $W_{B,i} B_{Y,i} W_{B,i}^* = B_{X,i}$. Now J_i and $W_{B,i} J_i W_{B,i}^{-1}$ are unipotent elements of $\mathcal{C}(B_{X,i})$, so using Algorithm 4 we can compute $W_{J,i} \in \mathcal{C}(B_{X,i})$ such that $W_{J,i} W_{B,i} J_i W_{B,i}^{-1} W_{J,i}^{-1} = J_i$. Now take $W_i = W_{J,i} W_{B,i}$.
- $f_i \in \Phi_2$. This case is identical to the Φ_2 case in Step 2 of Sect. 6.4, where we never used the fact that J_i was semisimple.
- $f_i \in \Phi_3$. Following the Φ_3 case in Step 2 of the semisimple case in Sect. 6.4, we work in $\mathrm{GL}_{m_i}(E)$, with $E = F[t]/(f_i)$ (or $\mathbb{F}_q[t]/(f_i \bar{f_i})$ in the unitary case). We compute $\widetilde{H}_{X,i}$ and $\widetilde{H}_{Y,i}$, matrices of hermitian forms preserved by $\widetilde{J}_i \in \mathrm{GL}_{m_i}(E)$ where J_i is the embedding of $\widetilde{J}_i$ into $\mathrm{GL}_{m_i d_i}(F)$. As in the case $f_i \in \Phi_1$, we can suppose that $\widetilde{J}_i$ is unipotent, since scalar factors do not affect the computation. The construction of $\widetilde{W}_i \in C_{\mathrm{GL}_{m_i}(E)}(\widetilde{J}_i)$ such that $\widetilde{W}_i \widetilde{H}_{Y,i} \widetilde{W}_i^\dagger = \widetilde{H}_{X,i}$ is identical to that described above in the case $f_i \in \Phi_1$.

The procedure also works for orthogonal groups $\mathcal{C} = \mathcal{C}(Q)$ in characteristic 2 for the reasons stated just before Sect. 6.4.1.

7.4.1 Special and Omega Groups

Here we assume that X and Y are conjugate in $\mathcal{S}$ or Ω. As above, we can compute a conjugating element $Z = P_X^{-1} W P_Y$ in the isometry group $\mathcal{C}$. If Z has inappropriate determinant or spinor norm, then we adjust Z to obtain a conjugating element in $\mathcal{S}$ or Ω. As in the semisimple case, we aim to compute $D \in C_{\mathcal{C}(B_X)}(J)$ having the appropriate determinant or spinor norm, and then replace Z by $P_X^{-1} D W P_Y$.

For $i = 1, \ldots, h$, let m_{ij} be as defined at the beginning of Sect. 7.2.

Suppose first that $\mathscr{S}$ is the special orthogonal group. If X and Y are conjugate in $\mathscr{S}$, then by the proof of Proposition 7.1.2(ii) we may suppose that there is an elementary divisor $f_1^m = (t \pm 1)^m$ with m odd, so we can choose

$$D = \begin{pmatrix} D_1 & \\ & \mathbb{I} \end{pmatrix}, \tag{7.4.1}$$

where $D_1 \in C_{\mathcal{C}(B_{X,1})}(J_1)$ has the same determinant as Z.

Now let $\mathscr{S}$ be the special unitary group. Set

$$r_i = \gcd(q+1, m_{i1}, \ldots, m_{ik_i}),$$

and let $r = \gcd(q+1, r_1, \ldots, r_h)$. Let $\omega \in \mathbb{F}_{q^2}$ have order $q+1$. From the proof of Proposition 7.1.2(i), we can find $H_i \in C_{\mathcal{C}(B_{X,i})}(J_i)$ such that $\det(H_i) = \omega^{r_i}$. Compute integers a_i such that $\sum_{i=1}^{h} a_i r_i = r$ in $\mathbb{Z}_{q+1}$, and let

$$H = \begin{pmatrix} H_1^{a_1} & & \\ & \ddots & \\ & & H_h^{a_h} \end{pmatrix}.$$

Then $\det(H) = \omega^r$, and since X and Y are conjugate in $\mathscr{S}$, there exists an integer ℓ such that $\det(Z) = \det(H^\ell)$. So we take $D = H^{-\ell}$.

Finally, we consider the case where $X, Y \in \Omega$ are Ω-conjugate, where Ω is an orthogonal group. In even characteristic, in the notation of Proposition 7.1.4 we can suppose that $c_\Omega = 2$, so there exists $f_1 \in \Phi_1$ and $D_1 \in C_{\mathcal{C}(B_{X,1})}(J_1) \setminus \Omega(B_{X_1})$, and we choose

$$D = \begin{pmatrix} D_1 & \\ & \mathbb{I} \end{pmatrix}.$$

Now suppose we are in odd characteristic, and that X and Y are conjugate in Ω. From the special group case handled above, we can compute a conjugating element Z in $\mathscr{S}$. Assume that $C_{\mathscr{S}}(X) \neq C_\Omega(X)$, and that $Z \in \mathscr{S} \setminus \Omega$. Then, in the notation of Proposition 7.1.3, $c_\Omega = 2$, and one of the following applies.

- There exists $f_i \in \Phi_2 \cup \Phi_3$ and m_{ij} odd. Now $C_{\mathcal{C}(B_{X,i})}(J_i)$ contains elements of non-square spinor norm. Hence, we take

$$D = \begin{pmatrix} \mathbb{I} & & \\ & D_i & \\ & & \mathbb{I} \end{pmatrix},$$

  where $D_i \in C_{\mathcal{C}(B_{X,i})}(J_i)$ has the same spinor norm as $P_X^{-1} W P_Y$.

- There exists $f_i \in \Phi_1$ with $C_{\mathcal{S}(B_{X,i})}(J_i) \nsubseteq \Omega(B_{X,i})$. Proposition 7.1.3 shows that we can choose $D_i \in C_{\mathcal{S}(B_{X,i})}(J_i)$ with the same spinor norm as Z, so we take

$$D = \begin{pmatrix} \mathbb{I} & & \\ & D_i & \\ & & \mathbb{I} \end{pmatrix}.$$

- There exist $f_1, f_2 \in \Phi_1$ such that $C_{\mathcal{S}(B_{X,i})}(J_i) \subseteq \Omega(B_{X,i})$ and $C_{C(B_{X,i})}(J_i) \nsubseteq \mathcal{S}(B_{X,i})$ for $i = 1, 2$. Here we take

$$D = \begin{pmatrix} D_1 & & \\ & D_2 & \\ & & \mathbb{I} \end{pmatrix}$$

with $D_i \in C_{C(B_{X,i})}(J_i) \setminus C_{\mathcal{S}(B_{X,i})}(J_i)$ for $i = 1, 2$.

7.5 Number of Conjugacy Classes in Isometry Groups

We found it useful to check that the number of conjugacy class representatives for isometry groups agrees with the results of Macdonald [23] and Wall [32, §2.6 and Thm. 3.7.3].

Theorem 7.5.1

- *The number of conjugacy classes of* $\mathrm{GL}_n(q)$ *is the coefficient of* t^n *in the formal power series*

$$\prod_{i=1}^{\infty} \frac{1-t^i}{1-qt^i}.$$

- *The number of conjugacy classes of* $\mathrm{GU}_n(q)$ *is the coefficient of* t^n *in*

$$\prod_{i=1}^{\infty} \frac{1+t^i}{1-qt^i}.$$

- *The number of conjugacy classes of* $\mathrm{Sp}_n(q)$ *with* q *odd is the coefficient of* t^n *in*

$$\prod_{i=1}^{\infty} \frac{(1+t^{2i})^4}{1-qt^{2i}}.$$

- *Let q be odd. Let k_n^+ and k_n^- denote the numbers of conjugacy classes of $\mathrm{O}_n^+(q)$ and $\mathrm{O}_n^-(q)$ respectively (with $\mathrm{O}_n^+(q) = \mathrm{O}_n^-(q) = \mathrm{O}_n(q)$ if n is odd). Then*

$$\sum_{n=0}^{\infty}(k_n^+ + k_n^-)t^n = \prod_{i=1}^{\infty}\frac{(1+t^{2i-1})^4}{1-qt^{2i}}$$

$$\sum_{n=0}^{\infty}(k_n^+ - k_n^-)t^n = \prod_{i=1}^{\infty}\frac{1-t^{4i-2}}{1-qt^{4i}}.$$

- *Let q be even. Define a sequence of polynomials $\chi_i = \chi_i(a,b,t)$ and a power series $\chi = \chi(a,b,t)$ as follows:*

$$\begin{aligned}
\chi_{-1} &= a, \\
\chi_0 &= b, \\
\chi_{2m+1} - \chi_{2m} &= t^{2m+1}\chi_{2m-1}, \\
\chi_{2m+2} - \chi_{2m+1} &= t^{m+1}(1+t^{m+1})(\chi_{2m+1} + (1-t^{2m+1})\chi_{2m-1}), \\
\chi(a,b,t) &\equiv \chi_{2m} \mod t^m
\end{aligned}$$

for all $m \geq 0$. The number of conjugacy classes of $\mathrm{Sp}_{2m}(q)$ (resp. $\mathrm{O}_{2m}^{\epsilon}(q)$) is the coefficient of t^{2m} in the power series $s(t^2)$ (resp. $\omega_\epsilon(t^2)$), where

$$s(t^2) = \chi(0,1,t^2)\prod_{i=1}^{\infty}(1-qt^{2i})^{-1},$$

$$\omega_+(t^2) + \omega_-(t^2) = \chi(1,1,t^2)\prod_{i=1}^{\infty}(1-qt^{2i})^{-1},$$

$$\omega_+(t^2) - \omega_-(t^2) = \prod_{i=1}^{\infty}\frac{1-t^{4i-2}}{1-qt^{4i}}.$$

In Table 7.1 we record the resultant polynomials in q for the isometry groups of specified type and degree n at most 10.

Table 7.1 Polynomials for number of classes in isometry groups

Type	n	Polynomial
GL	2	$q^2 - 1$
GL	3	$q^3 - q$
GL	4	$q^4 - q$
GL	5	$q^5 - q^2 - q + 1$
GL	6	$q^6 - q^2$
GL	7	$q^7 - q^3 - q^2 + 1$
GL	8	$q^8 - q^3 - q^2 + q$
GL	9	$q^9 - q^4 - q^3 + q$
GL	10	$q^{10} - q^4 - q^3 + q$
GU	2	$q^2 + 2q + 1$
GU	3	$q^3 + 2q^2 + 3q + 2$
GU	4	$q^4 + 2q^3 + 4q^2 + 5q + 2$
GU	5	$q^5 + 2q^4 + 4q^3 + 7q^2 + 7q + 3$
GU	6	$q^6 + 2q^5 + 4q^4 + 8q^3 + 11q^2 + 10q + 4$
GU	7	$q^7 + 2q^6 + 4q^5 + 8q^4 + 13q^3 + 17q^2 + 14q + 5$
GU	8	$q^8 + 2q^7 + 4q^6 + 8q^5 + 14q^4 + 21q^3 + 25q^2 + 19q + 6$
GU	9	$q^9 + 2q^8 + 4q^7 + 8q^6 + 14q^5 + 23q^4 + 33q^3 + 36q^2 + 25q + 8$
GU	10	$q^{10} + 2q^9 + 4q^8 + 8q^7 + 14q^6 + 24q^5 + 37q^4 + 49q^3 + 50q^2 + 33q + 10$
Sp (q even)	2	$q + 1$
Sp (q even)	4	$q^2 + 2q + 3$
Sp (q even)	6	$q^3 + 2q^2 + 5q + 4$
Sp (q even)	8	$q^4 + 2q^3 + 6q^2 + 9q + 7$
Sp (q even)	10	$q^5 + 2q^4 + 6q^3 + 11q^2 + 16q + 10$
Sp (q odd)	2	$q + 4$
Sp (q odd)	4	$q^2 + 5q + 10$
Sp (q odd)	6	$q^3 + 5q^2 + 15q + 24$
Sp (q odd)	8	$q^4 + 5q^3 + 16q^2 + 39q + 51$
Sp (q odd)	10	$q^5 + 5q^4 + 16q^3 + 44q^2 + 90q + 100$
GO	3	$2q + 4$
GO	5	$2q^2 + 6q + 14$
GO	7	$2q^3 + 6q^2 + 20q + 28$
GO	9	$2q^4 + 6q^3 + 22q^2 + 48q + 62$
GO^+ (q even)	2	$1/2q + 1$
GO^+ (q even)	4	$1/2q^2 + 5/2q + 2$
GO^+ (q even)	6	$1/2q^3 + 2q^2 + 7/2q + 3$
GO^+ (q even)	8	$1/2q^4 + 2q^3 + 5q^2 + 8q + 7$

(continued)

Table 7.1 (continued)

Type	n	Polynomial
GO^+ (q even)	10	$1/2q^5 + 2q^4 + 9/2q^3 + 9q^2 + 13q + 9$
GO^+ (q odd)	2	$1/2q + 5/2$
GO^+ (q odd)	4	$1/2q^2 + 4q + 17/2$
GO^+ (q odd)	6	$1/2q^3 + 7/2q^2 + 23/2q + 37/2$
GO^+ (q odd)	8	$1/2q^4 + 7/2q^3 + 13q^2 + 63/2q + 85/2$
GO^+ (q odd)	10	$1/2q^5 + 7/2q^4 + 25/2q^3 + 34q^2 + 72q + 171/2$
GO^- (q even)	2	$1/2q + 2$
GO^- (q even)	4	$1/2q^2 + 3/2q + 2$
GO^- (q even)	6	$1/2q^3 + 2q^2 + 9/2q + 4$
GO^- (q even)	8	$1/2q^4 + 2q^3 + 4q^2 + 7q + 6$
GO^- (q even)	10	$1/2q^5 + 2q^4 + 9/2q^3 + 10q^2 + 15q + 10$
GO^- (q odd)	2	$1/2q + 7/2$
GO^- (q odd)	4	$1/2q^2 + 3q + 17/2$
GO^- (q odd)	6	$1/2q^3 + 7/2q^2 + 25/2q + 39/2$
GO^- (q odd)	8	$1/2q^4 + 7/2q^3 + 12q^2 + 61/2q + 83/2$
GO^- (q odd)	10	$1/2q^5 + 7/2q^4 + 25/2q^3 + 35q^2 + 74q + 173/2$

Bibliography

1. Adams, W.W., Loustaunau, P.: An Introduction to Gröbner Bases. Graduate Studies in Mathematics, vol. 3. American Mathematical Society, Providence (1994)
2. Bäärnhielm, H., Holt, D.F., Leedham-Green, C.R., O'Brien, E.A.: A practical model for computation with matrix groups. J. Symbolic Comput. **68**, 27–60 (2015)
3. Bosma, W., Cannon, J., Playoust, C.: The Magma algebra system. I. The user language. J. Symbolic Comput. **24**, 235–265 (1997)
4. Britnell, J.R.: Cycle index methods for matrix groups over finite fields. DPhil. Thesis, University of Oxford, 2003
5. Burness, T.C., Giudici, M.: Classical groups, derangements and primes. Australian Mathematical Society Lecture Series, vol. 25. Cambridge University Press, Cambridge (2016)
6. Cannon, J.J., Holt, D.F.: Computing conjugacy class representatives in permutation groups. J. Algebra **300**, 213–222 (2006)
7. Carter, R.W.: Centralizers of semisimple elements in the finite classical groups. Proc. Lond. Math. Soc. **42**, 1–41 (1981)
8. Clarke, M.C., Premet, A.: The Hesselink stratification of nullcones and base change. Invent. Math. **191**, 631–669 (2013)
9. De Franceschi, G.: Centralizers and conjugacy classes in finite classical groups. Ph.D. Thesis, University of Auckland, 2018. researchspace.auckland.ac.nz/handle/2292/45197
10. De Franceschi, G., Liebeck, M.W., O'Brien, E.A.: Conjugacy classes in finite classical groups. github.com/eamonnaobrien/ClassicalConjugacy
11. Dietrich, H., Leedham-Green, C.R., O'Brien, E.A.: Effective black-box constructive recognition of classical groups. J. Algebra **421**, 460–492 (2015)
12. Fong, P., Srinivasan, B.: The blocks of finite classical groups. J. Reine Angew. Math. **396**, 122–191 (1989)
13. von zur Gathen, J., Gerhard, J.: Modern Computer Algebra. Cambridge University Press, Cambridge (2003)
14. Gonshaw, S., Liebeck, M.W., O'Brien, E.A.: Unipotent class representatives for finite classical groups. J. Group Theory **20**, 505–525 (2017)
15. Grove, L.C.: Classical Groups and Geometric Algebra. Graduate Studies in Mathematics, vol. 39. American Mathematical Society, Providence (2002)
16. Herstein, I.N.: Topics in Algebra, 2nd edn. Xerox College Publishing, Lexington/Toronto (1975)
17. Holt, D.F., Eick, B., O'Brien, E.A.: Handbook of Computational Group Theory. Discrete Mathematics and Its Applications. Chapman & Hall/CRC Press, Boca Raton (2005)

G. De Franceschi et al., *Conjugacy in Finite Classical Groups*, Springer Monographs in Mathematics, https://doi.org/10.1007/978-3-031-86461-2

18. Hulpke, A.: Conjugacy classes in finite permutation groups via homomorphic images. Math. Comput. **69**, 1633–1651 (2000)
19. Hulpke, A.: Computing conjugacy classes of elements in matrix groups. J. Algebra **387**, 268–286 (2013)
20. Huppert, B.: Isometrien von Vektorräumen 1. Arch. Math. (Basel) **35**, 164–176 (1980)
21. Kleidman, P., Liebeck, M.: The Subgroup Structure of the Finite Classical Groups. London Mathematical Society Lecture Note Series, vol. 129. Cambridge University Press, Cambridge (1990)
22. Liebeck, M.W., Seitz, G.M.: Unipotent and Nilpotent Classes in Simple Algebraic Groups and Lie Algebras. Mathematical Surveys and Monographs, vol. 180. American Mathematical Society, Providence (2012)
23. Macdonald, I.G.: Numbers of conjugacy classes in some finite classical groups. Bull. Aust. Math. Soc. **23**, 23–48 (1981)
24. Macdonald, I.G.: Symmetric functions and Hall Polynomials, 3nd edn. Oxford Mathematical Monographs. Clarendon Press/Oxford University Press, New York (1995)
25. Milnor, J.: On isometries of inner product spaces. Invent. Math. **8**, 83–97 (1969)
26. Murray, S.H.: Conjugacy classes in maximal parabolic subgroups of general linear groups. J. Algebra **233**, 135–155 (2000)
27. Neunhöffer, M., Praeger, C.E.: Computing minimal polynomials of matrices. LMS J. Comput. Math. **11**, 252–279 (2008)
28. Rylands, L.J., Taylor, D.E.: Matrix generators for the orthogonal groups. J. Symb. Comput. **25**, 351–360 (1998)
29. Steel, A.: A new algorithm for the computation of canonical forms of matrices over fields. J. Symb. Comput. **24**, 409–432 (1997)
30. Taylor, D.E.: The Geometry of the Classical Groups. Sigma Series in Pure Mathematics, vol. 9. Heldermann Verlag, Berlin (1992)
31. Taylor, D.E.: Pairs of generators for matrix groups, I (2022). arXiv:2201.09155
32. Wall, G.E.: On the conjugacy classes in the unitary, symplectic and orthogonal groups. J. Aust. Math. Soc. **3**, 1–62 (1963)
33. Wall, G.E.: The semisimple conjugacy classes in the symplectic groups (2015). arXiv:1512.04520
34. Wilson, J.B.: Optimal algorithms of Gram-Schmidt type. Linear Algebra Appl. **438**, 4573–4583 (2013)

Index

G. De Franceschi et al., *Conjugacy in Finite Classical Groups*, Springer Monographs in Mathematics, https://doi.org/10.1007/978-3-031-86461-2

The manufacturer's authorised representative in the EU is Springer Nature Customer Service Centre GmbH, Europaplatz 3, 69115 Heidelberg, Germany. If you have any concerns regarding our products, please contact ProductSafety@springernature.com

Printed and bound by CPI Group (UK) Ltd, Croydon, CR0 4YY
30/09/2026
02225595-0020